Einstein's Violin

Jeremy P. Tarcher/Penguin

a member of Penguin Group (USA) Inc.

New York

Einstein's
Violin

A Conductor's Notes

on Music, Physics,

and Social Change

Joseph Eger

JEREMY P. TARCHER/PENGUIN
Published by the Penguin Group
Penguin Group (USA) Inc., 375 Hudson Street, New York, New York 10014, USA • Penguin Group
(Canada), 10 Alcorn Avenue, Toronto, Ontario M4V 3B2, Canada (a division of Pearson Penguin Canada
Inc.) • Penguin Books Ltd, 80 Strand, London WC2R 0RL, England • Penguin Ireland, 25 St Stephen's
Green, Dublin 2, Ireland (a division of Penguin Books Ltd) • Penguin Group (Australia), 250 Camberwell
Road, Camberwell, Victoria 3124, Australia (a division of Pearson Australia Group Pty Ltd) • Penguin Books
India Pvt Ltd, 11 Community Centre, Panchsheel Park, New Delhi–110 017, India • Penguin Group (NZ),
Cnr Airborne and Rosedale Roads, Albany, Auckland 1310, New Zealand (a division of Pearson New
Zealand Ltd) • Penguin Books (South Africa) (Pty) Ltd, 24 Sturdee Avenue, Rosebank, Johannesburg 2196,
South Africa • Penguin Books Ltd, Registered Offices: 80 Strand, London WC2R 0RL, England

Most Tarcher/Penguin books are available at special quantity discounts for bulk purchase for sales
promotions, premiums, fund-raising, and educational needs. Special books or book excerpts
also can be created to fit specific needs. For details, write Penguin Group (USA) Inc.
Special Markets, 375 Hudson Street, New York, NY 10014.

Library of Congress Cataloging-in-Publication Data

Eger, Joseph, date.
    Einstein's violin : a conductor's notes on music, physics, and social change / Joseph Eger.
      p.   cm.
    ISBN 1-58542-388-2 (alk. paper)
    1. Music—Philosophy and aesthetics.   I. Einstein, Albert, 1879–1955.   II. Title.

ML3800.E3   2005   2004059817
781'.1—dc22

Printed in the United States of America
10 9 8 7 6 5 4 3 2 1

This book is printed on acid-free paper. ♾

Book design by Kate Nichols

This book is dedicated to all young people,
and to young-thinking rebels of all ages,
who struggle to make the world a better place.

Acknowledgments

The inspiration for this book came from so many sources that it would take another book to list them. Let's see: There were the B's Beethoven, Bach, Brahms, Berlioz, and the Beatles though my musical alphabet doesn't stop there; it goes on to many musical genres, historical periods, performers, and teachers.

My pantheon of scientists probably starts with the sheepherder who, night after night, looked up at the sky and began to speculate about how the world and stars interrelate. Then there are the hundreds of scientists who unknowingly cheered me on, long before the thought of writing a book entered my mind. Here's a short list: Pythagoras, Plato, Galileo, Newton, Descartes, Planck, and the Curies. In the twentieth century there was David Bohm (the first to encourage me in my "theory" that the universe is made of music), Einstein (my hero), Bohr, Heisenberg,

Szilard, Pauli, Oppenheimer, Feynman, Greene, Witten, and the entire alphabet of those who have dedicated their lives to the understanding of how things work.

I also acknowledge the majority of musicians, scientists, and millions of ordinary people who are aware of things *not* working, and who continue to struggle to improve and change conditions toward achieving a more rational and just society.

I'm afraid you'll have to read my book to get an idea of all those musicians, scientists, economists, public servants, philosophers, even those connected with government whom I single out and to whom I give special attention for their role in social change.

As for social change, it started with my experiences as a small child when I realized that there was injustice, ignorance, and cruelty all about me. I'm deeply indebted to all those who tried to do something about this and to help create a more just, sane, and decent society. It's a long and noble list.

More specifically, I'm indebted to those who read, responded to, and helped edit bits or all of the book's various incarnations. Among these are Dr. Ralph A. Dale, writer, musician, scientist, and lifetime fighter for rational social change, who forced me to think rigorously about premises, style, content, and continuity. Phyllis Kahaney, Tiffany Anderson, Mark McDiarmid, Christopher Atkinson, Marilyn Miller, Elly Friedman, Linda Ramirez, Billie and Philip Kirpich, Ralph Singer, and others read bits and criticized.

Of very special note is Mitch Horowitz, my editor extraordinaire. Mitch patiently and impatiently guided me through the shoals and trauma of a first-time book writer. He never lost confidence even when I did. Another of the remarkable and helpful team at Penguin is my copy editor, Mark Birkey. I also thank public-relations executives Ken Siman and Kat Kimball and sales director Fred Huber.

In a class by herself is my wife, Dorita Beh Eger, a loving support system, a cheerleader, and, as an excellent writer herself, a constant and valuable critic.

My everlasting gratitude to all of these and my apologies to those I'm sure I left out. Blame it on an extended senior moment.

Joseph Eger, 2005

Contents

Preface

*"Only art and science give us intimations and hopes
of a higher life."* —LUDWIG VAN BEETHOVEN

The universe is full of mysteries. Some, like absolute belief that the earth is flat, were explained at a leisurely pace over many centuries. Then came the scientific explosion of the twentieth century when Einstein triggered a century-long revolution of how we think about the world. His relativity theories turned our "good sense" ideas about space and time, energy and mass, and stars and galaxies upside down. He was soon joined by many other scientists, especially Niels Bohr, Werner Heisenberg, and Edward Witten, who were in the vanguard of "weird" quantum and string theories that turned our strongly held notions about atoms inside out. Nuclear energy, the atomic bomb, and a theory of dimensions of ten or more were among the results. It was a century that was very kind to writers of science fiction.

Those involved in the revolution were far from united. Controversy ran rampant, even challenging relativity's edict that nothing can exceed the speed of light. But controversy in science, and incidentally in music, were pale reflections of those in an increasingly global society. Wars, destruction of the environment, and the killing of people on massive scales threatened any kind of stability into the twenty-first century. Norms of ethical behavior, simple morals and compassion, even of laws and the U.S. Constitution were violated with impunity.

Then, in 2004, a backlash! People's movements sprang up, not without violence. Leaders emerged, with new weapons—computer, print, and film—with artists in the forefront. This book is an attempt to capture some of the salience of this battle for the soul and future of humankind.

1

How I Came to

Write This Book

"It is required of a man that he should share the passion and action of his time at peril of being judged not to have lived."

—OLIVER WENDELL HOLMES

M y life has been suffused with music and social concerns. Physics has been a longtime extracurricular interest. The social concerns hit me hard before I was seven years old, beginning within the framework of my family and expanding over the years to the ends of the earth.

Music arrived a few years later, and when it entered my life, it became my first and lasting love, eventually turning into a lifetime profession.

As for physics, I must confess to you early on that I'm not a physicist nor have I ever studied physics formally except for the old-fashioned grade school variety. But before being accused of false modesty, I will also admit that my bedtime reading has frequently included books on relativity and quantum mechanics. These subjects fascinated me because they were right in line with my feelings about music and with my lifelong social activism.

For me, music, physics, and social concerns are intertwined tightly together like a Navajo rug. The warp and woof of this book is woven from these three threads into patterns illuminating the effects of each on the other.

More than half a century ago, I came to the tentative conclusion that the universe is made of music. It took me another decade to work up the courage to propose this idea to Nobel Prize–winning physicist David Bohm. When he, followed by other prominent physicists, took my hunch seriously and gave it the lofty moniker "theory," I launched into further readings on relativity and quantum mechanics. I began to acquire more ideas about the relationships between music, physics, and society.

Confirmation arrived again when superstring theory came on the scene. One scientist after another used music as simile: A superstring is "like a violin string" vibrating in varied patterns, the tiniest strings as the smallest particles in the universe and the largest explaining the cosmos itself.

The following pages compose a conductor's notes on my seventy-year romance with music as it has intersected with my thoughts about physics, with my joys and sorrows, illnesses and healings, stumblings and enlightenments, and with my friendships and encounters with some well-known personalities in all three disciplines. I have learned more during the process of collecting these notes late in life than from all the other reading I have ever done. My reason for sharing them with you is my hope that you might become as enthusiastic as I am about the new, expanding opportunities for music and science as bridges to a more human and humane world.

Prelude: Theme and Variations

People are always surprised when I tell them that Albert Einstein was never without his violin; most don't even know he played the violin. He loved music and managed to practice every day, even when busily engaged in scientific work. When he was young, however, he found practicing scales and arpeggios tedious, until at age thirteen he discovered Mozart sonatas:

> *"My wish to reproduce them in their artistic content and their unique gracefulness forced me to improve my technique; this I acquired without practicing systematically. I believe that love is a better teacher than a sense of duty—at least for me."* —EINSTEIN

Musicians the world over are not too different. Except for verbal language, style, and technique, those I've worked with—whether professionals, pop stars, or amateurs like Einstein—in Moscow or Miami, Costa Rica or China, Bucharest or Boston, New York or Los Angeles, all seemed like one big family. There is something inherent in music that transcends differences in geography, culture, and even personality. The frequent experience of "good" music (what is good is explored later) in any genre, whether as listener or performer, seems to bring out special qualities of sensitivity and insight. Neurologists such as Richard Restak of George Washington University (author of the bestselling book *Mozart's Brain and the Fighter Pilot*) contend that music enhances brain circuitry and sharpens the brain's creative powers, stimulating dialogue between both sides of the brain. The mysterious magic and power of music also seem to infuse the entire plant and animal life. Nor does music stop there. As revealed by contemporary physics, music represents the oneness of all phenomena. All this and the implications for human society provided the seeds for this book. These seeds were well nourished and sprouted by Ludwig van Beethoven. Though born a couple of centuries apart, he and Einstein were my inspiration in writing it.

For Einstein, music was a kind of physics and physics a kind of music. He saw no dividing line and often said that science is an art. Beethoven's music too was composed of both disciplines. Above all, he and Einstein shared the same urge to reach for "a higher life."

It is this aim for a "higher life," a better life for the "soul" of all humankind that provides the book's building blocks. It defines and presages my definition of music. For me, "higher life" and "soul" are not abstract, utopian ideas; nor are they the saccharine, soupy-sentimental, Sunday-go-to-meetin' soul, or the superficial spins of moon/June tunesmiths.

Music goes far beyond the notes on a page or sounds on a horn, and, as can be described by twentieth-century physics, music is all-pervasive, saturating everything and everybody, from the tiniest particle to the entire universe.

The Power of Music and Science

My many musical and extramusical experiences have dramatically demonstrated music's power. Equally rife are the many exciting stories about the power of science. Both are intertwined with social change and process.

In the process of research for this book, I clarified and changed some former thoughts about the three subjects and came to conclusions other than the beliefs with which I began. For example, the "big bang" concept of a single universe no longer made sense to me. Nor did absolute and permanent music. Instead, I began to think in terms of "process." Process thinking is just a term for viewing things in their constant concatenation and development; however, I couldn't have arrived at that stage without traveling along traditional paths.

The march of music and physics through their historical and contemporary context led to my views concerning their potential role toward bringing about constructive change. The symphony orchestra gave me an apt metaphor, indeed an inimitable model, for society and possibly for the entire universe. Only partly whimsical was my metaphor for the universe as a cosmic orchestra; it has strings (string theory), brass (gravity and "dark matter" [or windy speeches!]), a composer/conductor (Nature or God), and percussion (which may be looked on as the rhythm in all life-forms or, if you prefer, the drumbeats of joy and tragedy). As in any orchestra, the resulting music interweaves among categories to fashion, in a word, music.

The physical theories about music's magic, established by philosopher-scientists such as Pythagoras thousands of years ago, provided fertile ground for the implausible new physics that burst upon the twentieth century, upsetting practically everything we thought we knew about reality.

My personal reality began early in the twentieth century as I observed a malfunctioning world all about me. From childhood on, inchoate yet overpowering feelings told me that something was wrong with much of the world and that I had the responsibility to set things back on course. Music slowly became my vehicle toward that goal while at the same time serving as balm to my troubled inner world. Later, science confirmed what I had long intuited about social change.

Variations on a Theme:
Cosmic and Earthly Music

I began reading up on relativity theories and quantum mechanics in order to better understand the relationships between music, physics, and a wonderful world, though the world was turning sour, indeed, the stench was becoming so bad that I could not remain uninvolved. The need for employing the powerful vehicles of music and physics for constructive social change became more pressing than ever. At first my involvement was chiefly by way of my music and the many musical events that others and I produced, but I soon came to realize that I needed a rational theory to back up my activism.

Unreality

For centuries we have been living in an unreality that has finally come to a head. Today's crisis goes well beyond music and physics, threatening society as a whole. One approach to understanding the problem is to look at the world in two different ways:

1. We can view the world by way of the old paradigms in science and society such as a physics of fragmentation—which we can label "Cartesian" (from Descartes) or "Newtonian," with their concomitant societal separations by race, religion, nation-states, etc. This outdated approach continues to guide our leaders as they separate the world into "them" against "us," nation versus nation, race against race, and religion against religion. Rich against poor has reached new heights of flagrant egregiousness, bolstered by a convenient dichotomy—"terrorists" against the good guys. "Terrorism" has become a slippery and flexible category, depending on who defines the bad guy of the moment.

2. A newer view of the world has been provided by the modern physics of relativity and quantum theories. For over a century, these and a scientific view of society have given us maps to help guide us to the reality of an unimaginably vast cosmos and a mind-bogglingly, tiny micro-universe. These point to the unity of all phenomena at all levels. Physics has proven the obsolescence of thinking of the fundamental nature of matter in terms of solid building blocks or even unchanging particles. Instead, the fundamental nature of the material world is best viewed through the underlying patterns of ever changing waves and waveforms. These patterns or geometric structures of form and proportion exist throughout nature, indeed in the entire universe. Our artifacts reflect these forms, in our homes, cathedrals, Indian rugs, and modern art. All of us are composed of vibratory waveforms. You and I are nothing but differently configured frequencies!

The implications of the new approach are far-reaching. Process thinking, if grasped in all its potential, can direct us to a brave new world. The how, as seen through the eyes of a symphony conductor and lifelong musician, supplies the grist for these "notes."

"Hearing the Echo of Earthly Music . . . Science Again Discovers How Much We Resemble the Rest of Creation"
—THE NEW YORK TIMES, January 17, 2001

My notes are not directed toward physicists, first of all because advanced mathematics (a must for physicists) is, well, "advanced." Very. Nor are they directed toward gray-bearded musicologists. I would rather reach those who are not yet stuck in obsolete patterns of Newtonian thinking and behaving, people who would like to participate in changing our world for the better (beards not excluded), toward a more harmonious world.

My many experiences in music of all genres over a period of seventy-plus years have convinced me of the power of music, far beyond the ordinary conception. What lay behind this power was opened up for me by the new physics and the chance to talk with physicists, either in person or secondhand by way of many exciting scientific books and articles. All this not only bolstered my theory that the universe is made of music but that the universe itself *is* music as it expands and contracts like the waves emanating from a plucked guitar string. My days have been enlightened with a wealth of information, emotions, and new friends while my life in music continued to bring joy, tragedy, excitement, and constant adventure.

2

Guns in My Face:

Music at My Back

Driving with a companion along a dusty, deserted road in the West Bank during the late seventies, I was suddenly faced with a platoon of Israeli soldiers pointing Uzi machine guns directly at my face. The commander approached my car, gun lifted, and demanded, as I rolled down my window, "Who are you and what are you doing here?"

"I am Joseph Eger. We are on our way to Amman to visit Queen Noor al Hussein." (How's that for name-dropping!) "We have an invitation."

He recognized my name, lowered his gun, and became friendly: "Aren't you the conductor?" starting a conversation about music, telling me about his uncle, a singer, and asking if I would be willing to audition him. He then tried to dissuade my companion and me from going into dangerous territory; the border soldiers would not let us through anyway, he said, and everybody would be leaving for the *Shabat* (Friday evening Sabbath) services. Seeing that we were determined, he motioned his troops to desist and called ahead to the Israeli military outpost to cooperate with us if they could. Music had established an immediate bond.

My companion and I had started out from Tel Aviv that Friday morning after four days of invigorating peace meetings chaired by Supreme Court Justice Chaim Cohen of Israel. We headed toward Jerusalem on our way to Jordan, where I had been invited to be a guest of the royal family. On arrival in Jerusalem, we stopped at a travel agency to get directions and make a phone call to Amman to let them know we were on our way. The people in the agency looked on us naive Americans as if we were crazy: "You'd better call your American Ambassador."

When we reached the ambassador, he patiently explained that the two countries were at war and that there were no means, by telephone or otherwise, of communication between the countries. Anyway, he cautioned, the Israelis would not let us go past the border.

Not to be discouraged, we returned to our car and headed toward the border. We had just passed through a hot, dusty Arab village when, a few miles down the deserted road, we suddenly faced those Uzi guns.

"Enemy" Territory

After persuading the Israeli colonel to let us through, and after his telling us that the border soldiers would not let us through anyway, we plowed ahead and soon reached a huge but almost deserted Israeli border outpost on the Jordan River. Two young soldiers greeted these two crazy Americans with amusement. They pointed out that everyone was leaving for *Shabat* and there was no place to leave our car and the Jordanians would not let us through anyway. We continued to wheedle and insist until one of the soldiers finally relented: "Well maybe my uncle will take care of your car, but we are sure you'll be back soon anyway." They magically summoned a Jordanian Volkswagen van.

More Guns

We entered the small van to face more guns, this time from a Jordanian officer and soldier. Both were dour and uncommunicative, though there was little time for talk since it took little more than a minute or two to cross the Allenby Bridge over the narrow Jordan River.

Now in Jordan, the colonel brusquely motioned to us to get out and directed us to a tiny military hut manned by one soldier with his gun ominously ready. The officer motioned for us to sit across the table from the soldier. We sat down as the officer and the Volkswagen soldier disappeared, presumably, we surmised, to get in touch with and get orders from the royal family about the claim of these foreigners in regard to an unlikely invitation from the queen.

We were sitting there for some time in absolute silence until the soldier, getting bored, turned on his small transistor radio. Music poured forth. My companion, thinking she recognized the music, turned to me and asked, "Isn't that Fayrouz?"

Music Opens the Door

Here I must give some background. Shortly after a series of my articles on Middle East peace appeared across the country, the Arab American Cultural Foundation, realizing that I was not an inveterate Arab hater, asked me to visit Lebanon to meet with the most popular musical star in the Arab world and, if I found her worthy, bring her to the U.S. for a tour. Even the Israelis enjoyed her music. The foundation sent me to Beirut, Lebanon, where I met this famous Fayrouz. After hearing her music, I invited her to the States. She readily agreed, and I subsequently became her music director for sold-out concerts at Kennedy Center, Carnegie Hall, in the United Nations. At the UN, her troupe of seventy dancers and singers joined my Symphony for United Nations orchestra in musical exchanges and performances. For that performance, I arranged, with difficulty (the Arabic musical scale is different from ours), the works in which our two groups collaborated. The international audience of diplomats, UN personnel, and invited VIPs applauded enthusiastically, for was not that what the UN was about? The Arab delegates beamed with pride.

Now let us fast-forward to the Jordanian military hut.

Seeing that we were acquainted with Fayrouz's music, the soldier dropped his gun to the table, perked up, and in halting English, "You . . . like . . . Fayrouz?"

We nodded our heads, yes.

The soldier had no idea that I actually *knew* Fayrouz and had been her music director. It was enough that we liked her music. His face became wreathed in smiles. To put it in American vernacular, we immediately became soul brothers.

When the commander returned to the hut, the soldier spoke to him in Arabic, and now his formerly sour face was also transformed. He became friendly and cooperative, though he regretfully informed us that he could not reach an official in the palace and so had no authorization to let us continue that day.

But the moral of the story is clear; music had come to the rescue with its powers to cross the artificial barriers that keep peoples apart.

Denouement

On my return to New York, I received a call from the Jordanian embassy passing along the message that the queen was very embarrassed—she and the family had been on the traditional long midday dinner break—and she offered me a complimentary first-class transoceanic ticket on a Jordanian Airlines jet at any time of my convenience, a luxurious trip that I subsequently enjoyed on my next trip to the Middle East, which included a most interesting week in Amman as a guest of the royal family. Queen Noor and her husband, the king, were most gracious.

In an earlier trip to the Middle East during the seventies, I honored an invitation from the Haifa Symphony Orchestra to guest conduct. At one of my rehearsals, a violinist in the orchestra walked up to the podium during intermission and offered: "Maestro, I'm going to the West Bank for a few days after the concert. Would you like to come along?"

"Where's the West Bank?" I asked.

In those days, such ignorance was common. To most Americans, the Middle East and the Arab world in general made up just one dark blob. We did not differentiate among the varied Arab countries and in general had little idea of the area's geography. I held off my decision until I could seek advice from some of my Israeli friends. A typical warning came from one Israeli *chaver* (friend):

"Don't go. It's very dangerous. The Arabs might invite you into their tents for tea—they are a hospitable people—but as you leave you might find a knife in your back. They are like dogs, unpredictable."

These reactions only intrigued me. Being an adventurous sort, I agreed to go. My violinist friend, Joseph Abileah, had been born in Palestine during the period when many Palestinian Arabs and Jews lived in relative peace, some literally next door to each other. Joseph was a "peacenik" and had many Palestinian friends.

We started out by foot since there was no public transportation across the rather amorphous border between Israel and the West Bank, nor were there roads or paths. We walked through the fields and brush, passing a large Israeli military outpost, and soon came to a small Arab village. For three days we walked, going from town to town by foot, bus, or taxi. We stopped in Jericho, Nablus, and other villages, visiting people Joseph knew or just hanging out in cafés and restaurants. We saw no tents and didn't have the pleasure of knives in our backs. Instead, we met with many fine people: doctors and priests, people in the streets and in buses, intellectuals and shopkeepers. One "sweet shop" owner, noting that my friend was an Israeli Jew and myself an American Jew, refused payment from his "Hebrew brothers." This reaction was fairly typical of our consistently friendly encounters. On our last night, we stayed at a spare but pleasant Christian house run by nuns in Jerusalem.

Return to Normalcy

The first Israeli I saw on my return to the Israeli portion of Jerusalem was a good friend to whom I related my surprisingly positive experience. She immediately warned me not to speak to other Israelis about having good experiences with Arabs. Since I had to return to California soon after, I managed to avoid discussing it with other Israelis. But when I returned to Los Angeles and met with two of my dearest friends, excellent musicians who had been best man and woman for my first marriage, I felt totally free to respond to their questions about my recent trip to Israel. They wanted to know about my concerts, my programs, the audiences, and my experiences in general. After telling about my concerts, I began rather enthusiastically to recount my unexpectedly pleasant visit to the West Bank.

As I unfolded my tale, their faces grew darker and darker. My heart froze as I quickly trailed off to a diminuendo, shifting gears to another subject. This encounter taught me to be wary about raising the subject and to choose with extreme care the people with whom I would discuss the Middle East.

However, I began to read the newspapers with new eyes, discriminating between what appeared on the subject and my actual experiences on the ground. It became apparent to me that a mammoth mythology about the Middle East had been created in the news media. This seemed to be not only counterproductive in terms of any hopes for stability and peace in the area, but clearly, misinformation and actual lies were being spread. Unable to stand this disparity any longer, I put some thoughts on paper, called *Newsweek* editor David Alpern, who had once interviewed me, and asked if I could send him an article I had written. He replied that he would pass it along to the appropriate editors. They allotted me a full "My Turn" page in the September 15, 1980, issue. My article, titled "Is It Good for the Jews?," pointed out what is accepted wisdom today, that "a conflagration there would ignite the world," and that "the only path to peace and justice was dialogue and a Palestinian State." Also, I sermonized: "We (Israel and Diaspora Jews) may find, if it isn't too late, that what is good for the Arabs—and for all peoples—is also good for the Jews." My contention had a basis in the fact that, as a result of my music making with Arabs, I had developed a good relationship with the Palestinian Observer at the United Nations and with people close to Arafat. To a man, they promised that the Palestinians were ready for a peaceful accommodation. They repudiated the oft-repeated canard of "throwing the Israelis into the sea."

How very sad that what could have been accomplished with relative ease in those days, when music greased the wheels, has become a bloody cycle of death and destruction from policies that are clearly not working, either for the Jews or the Palestinians, escalating into a hell spreading throughout the Middle East and beyond. Finally, twenty-one years later, four former heads of the Shin Bet security service are saying that current Israeli policies are leading to catastrophe. Five hundred army officers and other men are refusing to fight against civilians, and a top minister in Sharon's government has come out strongly for negotiations. What music could have helped accomplish, politics destroyed. We can only hope it is not too late to revert to a musical model.

Oboes Versus Flutes

Before I became a conductor, I played solo first French horn in the New York Philharmonic, the Los Angeles Philharmonic, and the Israel Philharmonic. In the first two of these prestigious orchestras, the solo first flutist and solo first oboist, who customarily sit side by side in the center of the orchestra, just did not get along. To be more accurate, they hated each other.

For more than twenty-five years, these superb artists on flute and oboe, which instruments have key roles in all symphony orchestras, did not speak to each other! So what could they do? Let us imagine diabolical schemes they might concoct to vent their mutual dislike. How about one sabotaging the other by playing wrong notes? Or out of tune? Or not together? This would hardly be a solution, for they would succeed only in sabotaging themselves. So for all those years they managed to play together. Magnificently! After all, these four artists were also first-class professionals.

The moral is obvious: regardless of differences among people, enlightened self-interest can unite by providing a common goal, in these cases keeping their jobs and, above all, making splendid music. At least that is what Beethoven said in his Ninth Symphony, setting Friedrich Schiller's "Ode to Joy," with its line "Alle Menschen werden Bruder" ("All humankind *shall* be as brothers")—Beethoven musically emphasized the "shall."

3

How It All Began—
Back to 1951

W hen Leonard Bernstein, then conductor of the
New York Philharmonic, asked me to go to Israel to play first French horn and coach the winds and
brass players of the Israel Philharmonic, I was happy to
comply. The string players have always been excellent in
the IP, but the brass and woodwinds were not up to par,
because it had been only a few years since World War II
and the 1948 war for Israel's independence. In the interest of full disclosure, I must admit a guess that Lenny
also wanted to give me a respite from the Un-American
Activities Committee hearings and McCarthy witch hunt.

Those were dark days in American history. I was not
alone among those who had been snared by the committee. Included were hundreds of others in music,
science, film, education, and government. Especially targeted were well-known actors, directors, writers, pro-

ducers, musicians, and others in Hollywood and New York, as well as in universities across the nation. Albert Einstein, playwright Arthur Miller, and even President Eisenhower were among those tarred by being called "communists," "dupes of communists," or "fellow travelers." FBI Director J. Edgar Hoover directed special venom toward Charlie Chaplin, Einstein, and eventually John Lennon. Hoover had his agents build a huge file containing reports of Einstein's speeches and letters for peace and justice, trying to make them appear as evidence that Einstein was a disloyal American, a subversive, and a potential spy. It was hardly a secret that Einstein was ever active, starting with his hatred of Nazism in Germany, fascism in Italy and Spain, and injustice anywhere. He gladly lent his name to any honest organization fighting for these issues. Hoover made much of Einstein's pacifism and the physicist's belief that socialism and world federalism were the best ways to address society's problems.

"The world is a dangerous place to live in not because of those who do evil but because of those who watch and let it happen." —EINSTEIN

Einstein's example of a man who was at once scientist, musician, and principled citizen of the world gave me encouragement to stand up to the committee, though I must say that Israel and my musical duties proved to be a welcome breath of fresh air after the stench of repression and fear that had spread across the United States. While England and Europe looked on, ridiculing the antics of those crazy Americans, my experiences and those of other "blacklistees" were not so funny. For the subsequent twenty-plus years, I was shut out of one position after another. For one example: in 1971, unanimously selected and engaged by the Conductor Selection Committee of the San Diego Symphony after being recommended by some of the most prominent musicians of the century, I was rejected just as we were about to sign the contract. One member of the board was so excited by the prospect of having me there that he proposed donating a huge sum of money. At the last minute, another conductor who wanted the position presented a twenty-year-old clipping from the *Los Angeles Times* (1951) to the influential publisher of a right-wing newspaper chain. That clip included my name as one of the "unfriendly witnesses" before the House Un-American Activities Committee. My unfriendly testimony was as follows:

> Gentlemen, I play the French horn for a living. In order to play that difficult instrument, one has to keep one's head up. I'll be glad to tell you anything you want to know about myself, but I can't get others in trouble who never committed a crime. Should I learn that anyone has committed a crime, I would report that person to the proper authorities. I know of no such person. Thank you.

Shortly after that time, I lived in Israel for about a year, playing in the Israel Philharmonic and sharing living quarters in the Orchestra House with visiting concert artists such as Isaac Stern and other foreign musicians.

4

Fun in Music
and Physics

"If one hears bad music it is one's duty to drown it out by conversation." —OSCAR WILDE

It was great fun reading a variety of views about Einstein's relativity theories, the strange, science-fiction-like world of quantum mechanics, the four forces in the universe, the subnuclear microworld, and its macro cousin, cosmology. I came to relish my studies in physics almost as much as my studies in music. It is accepted wisdom that studying music equips young people and those not so young for virtually all other disciplines. Lifelong work in music gave me a big push for my latter-day studies in physics.

My student days at the Curtis Institute of Music in Philadelphia and in Tanglewood, Massachusetts, gave me an excellent musical education and the gift of fellow students who became friends for life. One or two helped radicalize this small-town naïf while contributing to my critical thinking in general. Those friends opened my mind to the world of progressive politics and made me, like students all over the world, a social activist involved in trying to change the world. No matter that financial insecurity was a way of life (budget of $0.35 a day for food and $3.50 per week for a tiny, fourth-floor garret with cold water five floors down in the basement). I did have some concern about pursuing a tenuous profession, especially after studying with a German teacher (during the early years of World War II) who made no bones about his Nazi sympathies, as he enjoyed baiting this only Jew in his class. But all this mattered little in the exhilarating atmosphere of music and friends.

> *"Never look at the trombones; it only encourages them."*
> —RICHARD STRAUSS

At one point, four of us lived in one room: besides myself, a cellist, who later joined the Boston Symphony, and a trombonist and a double bass player, both of whom graduated into the Philadelphia Orchestra. The bass player, Ed Arian, wrote a book or two about that orchestra, gained a doctorate or two in nonmusical subjects, and became head of the Pennsylvania State Council of the Arts. That left me, a French horn player. Imagine the cacophony when more than one of us practiced at a time! We had no kitchen, so we cooked our meals on a one-burner hot plate in the bathroom and washed our dishes in the bathtub. We were living Puccini's *La Bohème,* and, like the poor artists in that opera, we enjoyed our music and our camaraderie. We were musicians!

I continued as an orchestral musician, became a concert artist touring the world, and, in the last thirty-five years or so, have been a symphony conductor.

My student days began a lifetime quest toward finding ways to harness the mysterious power of music and related arts for social change. I roamed down that highway and its many byways, and ended up in the utterly fascinating fields of contemporary science. To my delight, I found that music, physics, and social change are inexorably locked arm in arm.

How I Became a Scientist
at Age Seven

My very first lesson on scientific thinking came to me when I was seven years old. I was the youngest of nine children of immigrant, Orthodox Jewish parents. When my oldest sister, Ethel, secretly married Ed French, suddenly everything changed. The sky fell in. Ed was not Jewish!

My parents "sat shivah," tantamount to excommunicating my sister. Mother covered all the mirrors with sheets. With a shawl over her head, she silently prayed before a candle. For days my father sat rigid and ashen-faced on a wooden crate, intoning prayers. Ethel was banished from our home and from all family connections.

Ed was a gentle man who worked in an aluminum plant in Parnassus, a tiny town in western Pennsylvania where my family lived. He truly loved my sister and would otherwise have been considered a good mate for one of six sisters who, in those times, should be "married off," especially as her marriage would mean one less mouth to feed.

At the turn of the twentieth century, my parents, Abraham Eger and Clara Ellovich, came to this country from a ghetto in Romania. Abe had no formal education outside that which all Orthodox men receive. His world was pretty much circumscribed by two things, making a living and being piously Jewish in an alien world. He was a good, hardworking man, faithful and devoted to his family. He labored and sacrificed to seek a foothold in the "new country." For him, the marriage of his daughter to a "goy" (gentile) was a major catastrophe. The result was pain for everyone in the family. My mother, Clara, had a few years of schooling in Romania. She struggled in the face of poor health and meager money to clean the house and cook kosher food for her large brood of ten. Going out to eat was unheard of. There were no books in the house, no pictures on the walls, no newspapers or magazines except one long-outdated and frayed *Reader's Digest*. Survival was the only game in our family. I felt for my parents and for Ethel, whose pain grew to mortal dimensions.

Soon after this family cataclysm, Ethel, as much mother to me as my overburdened mother, surreptitiously came to see me as I was playing on the street. Her palpable yearnings and hunger for her family and her baby brother were almost unbearable for me to witness. The shaft in my heart never quite healed. Nor did the one in hers.

When I saw the deep hurt in these adults who meant so much to me and to each other, I literally looked up to the heavens, shook my seven-year-old fist at the sky, and in my mind shouted: "If You exist, strike me dead!" How could it be that God and religion could do this thing? God gave no answers, and I became an instant skeptic. If I had known the words, I would have called myself agnostic or atheist, a challenger of entrenched prejudices. I no longer trusted adult thinking or dogma, religious or otherwise. I can still remember the very spot on the street in front of our house where this fist-shaking happened, and the turbulent feelings that accompanied it. How could my father, mother, and sister do such damage to their own? This adult world was not all it was cracked up to be.

From that moment, I became rebellious to any authority and converted to the skeptic's religion—science. The scientific approach to life's problems and promise became my mode of thinking. With the clarity of a child's mind, I reasoned that if religion, or at least Orthodoxy, can so damage good people, there must be something wrong with religion. Forsaking the God of my fathers gradually gave me a new God in the form of music and of science. Both gave me a sense of a more harmonious universe. Many years passed before I could discriminate among the many varieties of religious believers and belief systems.

I lost the innocence of childhood and became shy and introspective. It took many decades for me to regain a semblance of what it is to be a child, to relax and play, to experience amusement. The impact of this event and my outrage never quite left me. Through the years, I have continued to mentally shake my fist at the absurd, idiotic cruelty of so much of human behavior. I became a critic of sectarian beliefs and practices, of separation and fragmentation. Instead, I explored what I deemed to be other, more positive solutions. These lay in the fields of science and music. That seed germinated slowly, growing into a belief that political change was necessary. Music provided a way to express my jumbled, inchoate feelings. Science began to make sense of and give order to a world where I experienced disorder all about me.

Ed converted to Judaism, enduring the painful circumcision operation and the long Orthodox training required for conversion. The family became reconciled, but it was too late for Ethel. She became ill and died after five years of marriage. The cause was colitis, a disease that psychologists and psychiatrists link to emotional trauma. As this twenty-six-year-old woman lay on her deathbed in the hospital, she needed a blood transfusion. Doctors tested the blood of everyone in the family and found that I, now twelve years old, was the only one with the matching blood type. My blood did not save her, and she passed away in a matter of hours. I could not shake a feeling of responsibility for my failure and for her death. As she lay dying, *she* comforted *me*! Her loving good-bye in her last moments remains a vivid memory.

When I was fourteen or fifteen years old, my cousin Jackie and I were sitting on the grass in front of our school talking about what classes we intended to take in our sophomore year. He was a depressed and troubled boy. I was trying to cheer him up by sharing my excitement about life's possibilities. My vision of the world was highly imaginative. I told him how I saw connections between all people and all things. Since he was studying trombone to get into the band (which I had already joined), I talked about how every note we played traveled outside ourselves, and who knew how far? I expanded that idea to the complexity of interaction between musical notes, our breathing, our molecules, perhaps even our ideas. All of these phenomena were spreading out there somewhere in the world. Music was the deus ex machina for these speculations.

5

What It's Like

to Be a Musician

"Don't bother to look; I've composed all this already."
—GUSTAV MAHLER to Bruno Walter, who had stopped
to admire mountain scenery in rural Austria

"Musicians talk nothing but money and jobs. Give
me businessmen every time. They really are
interested in music and art."
—JEAN SIBELIUS, explaining why he rarely invited
musicians to his home

"I'm not handsome, but when women hear me play,
they come crawling to my feet." —NICCOLÒ PAGANINI

"Wagner's music is better than it sounds."
—MARK TWAIN

"Already too loud!"
—BRUNO WALTER, on seeing the players reach for their
instruments at his first rehearsal with an orchestra

I am proud to be a musician. Music is a noble art, but the profession is not always so noble. With the exception of a few at the top, it is very difficult to make a living at it. Of all the musicians listed in the musicians' unions of major cities, I would make an educated guess that well over 80 percent of these professionals are either unemployed or struggling to make a living as teachers or in some other calling. The title "professional musician" is a slippery designation.

Yet, everybody thinks they know what a professional musician is—a person who makes a living playing music. Most musicians do not make a living in their profession, and have to spend time in other pursuits. Amateurs keep hoping to become professionals, and mothers sacrifice to get lessons for their children. The conservatories grind out musicians like sausages.

Being in the business of music is like trying to smell roses at the bottom of an ill-smelling garbage can. Even while a musician is struggling to bring vivid emotion to life through his or her craft, he or she still has to deal with the realities of an extremely competitive and cutthroat society. Still, parents push their children to keep entering the field, and young people cannot resist the siren song of music's wonders. Why shouldn't they? All the peoples in the world venerate music of one kind or another. Music is irresistible.

The word "musician" covers a lot of territory, from Britney Spears and Mick Jagger to Wynton Marsalis and Chucho Valdez, to Yo-Yo Ma, Itzhak Perlman, and James Levine, with film and Broadway composers, rap and punk artists, folk singers, and a raft of others thrown in for good measure. Twentieth century "classical" composers, from Darius Milhaud and Igor Stravinsky to our contemporaries, use jazz and popular themes and rhythms as suits their purpose. The definition of a musician grows more abstruse. The word "serious" is sometimes used to describe "classical" music or musician, though that is perhaps even a less apt description. I know lots of pop, jazz, and rock musicians who are more serious than those of us who are called classical. My former instrument, the French horn, is usually associated with classical music, but a small group of adventurous souls, such as Julius Watkins, Willie Ruff, and a young John Varner have crossed the bridge rather successfully, as have string instrument players.

In 1975 Miles Davis declared jazz to be "the music of the museum." Today it is institutionalized, funded, and glorified in centers such as Lincoln Center and in universities around the country. Alfred Appel in his

Jazz Modernism: From Ellington and Armstrong to Matisse and Joyce, approaches jazz fairly comprehensively, providing the relationship with other contemporary arts.

I'm usually classified as a "classical musician," though I have performed concerts with jazz, pop, and rock groups, as well as improvisatory dance, theater, and the like, first as an orchestra player, then as a concert artist, and finally as a conductor with symphony orchestras. I conducted one of the first collaborations ever between classical music and rock, which involved groups such as the New York Rock Ensemble, the Rascals, Elephants Memory, John Lennon and Yoko Ono, and Keith Emerson of Emerson, Lake, and Palmer (ELP). *Rolling Stone* magazine called me "The Symphony Conductor We Can Trust" (whatever that means!). Note the quote marks around "classical musician." The classifications of music and musician are no longer clear-cut, as composers and performers everywhere cross over into their neighbor's territory.

6

What Is Music?

In the sixth century B.C.E., Pythagoras discovered that sound is subject to mathematical laws and proportions and suggested that other aspects of the world may be regulated by the same numbers: the seasons, the tides, and, yes, even the balance and discords of the human spirit. Music, which had acted as the binding force of society, was now seen to link all phenomena. In fact, there is some relationship between the Platonic postulate that music was the foundation and regulator of the universe and durable speculation about the "music of the spheres" in the universe (as popularized by the film *2001: A Space Odyssey*). By "music of the spheres," we mean the sounds of all heavenly bodies, as they move through the abyss of space, interacting with one another through gravity and other forces.

Pythagoras thought that music was the sound of mathematics and that numbers were the ultimate explanation of all things. There is a story of how Pythagoras discovered the link between number and harmony. He happened to pass by a brazier's shop, where he heard the hammers beating out a piece of iron on an anvil, producing sounds that harmonized. Entering the shop to investigate how this hammering could produce such harmonious sounds, he found that the musical intervals were in proportion to the ratio of the weights of the hammers.

He experimented by hanging different weights on strings of adjustable lengths, and plucked the strings so as to produce different sounds. He discovered that the most appealing sequences of musical tones were linked by simple arithmetical ratios of whole numbers that he and his followers revered. He discovered further that the shorter the string, the higher the note. Halving the length of a vibrating string produces a note that is higher by one octave; doubling its length produces a note that is one octave lower. The ear seems to like the combinations of notes produced by strings whose lengths are in the ratios 1:1, 1:2, 2:3, etc. The result of a ratio like 7:11 is discordant; still higher and odder ratios even more so.

The belief that numbers possess such meaning remained associated with the study of musical harmony for nearly two thousand years, affecting the religious reverence for these ratios. Plato took these up and later Aristotle, who argued that each of the celestial bodies must produce musical tones that will depend upon the distance of the body from earth and upon its speed. He said, "The motion of bodies of astronomical size must produce a noise, since on the earth the motion of bodies far inferior in size and speed of movement has that effect." Also, with the motion of the sun and the moon, and all the stars, so great in number and in size, how could they not produce a sound immensely great? Starting from this argument, and from the observation that their speeds, as measured by their distances, have the same ratio as do pleasing sounds, the ancient Greeks asserted that the circular movement of the stars is as sonic harmony. This led to the theory of the music of the spheres, still being studied in our universities. Shakespeare's Lorenzo describes this celestial harmony to Jessica just before the musicians enter in *The Merchant of Venice:*

> How sweet the moonlight sleeps upon this bank!
> Here will we sit, and let the sounds of music
> Creep in our ears. Soft stillness and the night
> Become the touches of sweet harmony;
> Sit Jessica, Look, how the floor of heaven
> Is thick inlaid with patens of bright gold
> There's not the smallest orb which thou behold'st
> But in his motion like an angel sings.

Early Music

The purpose of primitive music and early religious music was not entertainment, but was to assuage and praise the gods, help the hunt, improve the crops or the weather, frighten the enemy, arouse the troops, and signal for a thousand reasons. Some aspect of the arts accompanied the rights of puberty, death, and other ceremonies.

The court jesters and itinerant musicians were chroniclers, interpreters, and social critics, telling tales of other places and times or describing people and would-be heroes of the day. Songs were simple. A narrative was direct. Words meant what they meant. Language was used to communicate rather than to conceal, as is sometimes the case today.

Music was defined as the art of organizing tones to achieve an aesthetic experience. In ancient Greece, music was the union of word and sound as an inseparable unit sharing the same task: the expressing of an idea. Poetry and music were the transmitters and the mainstays of culture. For the Greeks, the first form of education was music, since it provided the introduction to laws and traditions. Music was considered necessary for the preservation of the community and was inextricably interwoven with the life of the state. All public occasions included music. There were contests at which singers and instrumentalists could win public acclaim.

Unlike visual art, numbers, or the written word, there has never been a culture without music. Musical instruments have been found in Cro-Magnon settlements in central and northwestern Europe. There are decorated flutes made from mammoth bones. Percussive instruments like castanets are between 20,000 and 29,000 years old. Artifacts indicate that these instruments were used in the performance of ceremonies. In ancient and modern civilizations, one could find music whenever there was a need to increase group bonding or inspire acts of courage.

The earliest, most spontaneous human-generated sounds may have been the cries of a baby. Perhaps these whimperings gave us sensitivity for particular sounds and developed a disposition toward musical sounds. The reaction of the mother in her soothing voice might have developed musical sounds. The biological conditioning begins antenatally, as the fetus responds to the body rhythms of the mother and to sounds outside the womb. The patterns of bird songs are capable of being quite elaborate. Some mimic our rhythmic patterns, and one, outside my window, sings a perfect arpeggio, time after time. I swear that when a note or two is a bit out of tune, the bird corrects it in the next rendition!

Charles Darwin tried to explain music by its possible origin in animal mating calls. We all know that the attempt to attract mates with music is not limited to animals. Where is the man or woman who can deny that they try to attract and seduce the opposite sex with the help of a little soft music!

Defining Music

In the attempt to define music, we run up against almost insurmountable problems. Is it atonalism, punk, rap, or the various versions of jazz? Is it the classic 12-tone compositions, Indian ragas, Middle Eastern scales that sound unlike any music we know? Is it Chinese or Japanese traditional music? Is it the experimental sounds that have been rife during the twentieth century, or is it all of these and much more?

How is the drop of a stone in the water different from the tap of a stick on timpani? Which is noise and which is sound and which is music? Famed musicologist Heinrich Schenker theorized that music is built from an inventory of notes and rules, with notes placed into a sequence and then organized into hierarchical structures, all superimposed on the same string of notes. Can this definition be reconciled with ultramodern music or the traditional music of Chinese opera? To understand a composition means to bring together these structures in our mind. Thus, this is a tyranny of notes superimposed on a musical instrument performed with the various timbres, pitches, and rhythms. Is music then a cultural construct, to be defined in each era by a particular culture, which finds various sound combinations pleasant or otherwise useful? Or is it strictly defined for all time by Pythagoras and the harmonic series as revealed so magnificently in Western classical music of previous centuries? A hornet's nest of questions!

The Power of Music and the Ancients

Most traditional cultures regard sound as sacred, as the manifestation of a primal rhythmic force that harmonizes the universe. Many cultures believe that drumming conveys the rhythms of seasons, tides, and moon cycles, as well as those of the human body.

Art forms such as Indian music and dance, Muslim calligraphy, and decorative arts and architecture consist of intricate patterns, numbers, and movement. In a word, wave frequencies. Each reflects cosmic order in its precision, beauty, and infinite creativity. A Gothic cathedral expresses the accumulated experience of many generations in the geometry of its architecture, its soaring arches and its stained-glass windows.

Nonverbal music has been extolled through the ages for its capacity to reach the innermost soul of man. From earliest times, man has known that there are psychological and emotional reactions to music. He used the powerful elements of rhythm and dynamics to incite or to tranquilize and to play on feelings and to influence mood. Primitive man used music for his healing rituals and to move the forces of nature to his bidding. In ancient Egypt, the musician/priestess was called upon to cure ills.

Egypt

Egypt developed a science of mysticism and a creation mythology. The Egyptian Book of the Dead asserts that God or his servant gods created everything by combining visualization with uttered sound. By pronouncing something's name, that thing would come into existence. A text from as late as 310 B.C.E. states, "Numerous are the forms from that which proceedeth from my mouth." The god Ra was also called Amen-Ra. The word *amen* has been equated with the Hindu *om*. Ra came into being by pronouncing his own name.

Egyptian accounts of the creation describe the primeval waters as a mass of energy. When the vibrations of the Word were poured into these waters, they created radiating currents and ripples with emerging complex patterns of geometrical forms. Thus was matter formed from the stress of "opposing" waves. Knowledge and skillful use of words of power could create and destroy form, heal the sick, and bring the dead back to life.

Egyptian music was deeply associated with mysticism. Like the Hebrews, the Egyptians believed that the creation was due to the gods of the seven tones, consisting of the pentatonic (five-tone) scale and two subsidiary tones. Translations of Egyptian hieroglyphics indicate that the Egyptian priesthood, by their "perfect and scientifically proportional speech" always showed the rays (tones) of the sun in such cosmic numbers as seven and twelve. The word for sound was *berw,* literally "voice" or "Word." Kabbalist Giulio Camillo of the sixteenth century claimed that priests were trained to employ guarded, proportional speech, causing the words to be "animated by a harmony." Since their scale was connected to the movement of the five known planets, the Egyptians may have had a major role in the teachings of Pythagoras and on the "harmony of the spheres."

The pentatonic system preserved the traces of primeval community, a sort of universally accepted impulse pointing toward an equally ancient separation of cultures. Yet it flowered in highly developed musical cultures. It may have been the first major link to advancing melody. The major scale must have had a strongly secular sound for it to make official defenders of classical scales of the late Middle Ages call it the *tonus lascivus* (which at the time meant "sweet" as well as "lascivious"!).

The Cure

The Hebrews also attributed curative powers to music. The Bible states of David and Saul: "And it came to pass when the evil spirit from God was upon Saul that David took a harp and played with his hand; so Saul was refreshed and was well, and the evil spirit departed from him."

In Greece, music was used to cure the outbursts of the mentally ill, and when disorders developed, the physician tried to restore the balance with medicine and music. Apollo was the god of both medicine and music.

When Philip V of Spain became ill and suffered from a very deep depression and was unable to carry on his affairs of state, his wife called the great castrato Carlo Broschi Farinelli to sing for him. The king immediately became well. He resumed his affairs of state and kept Farinelli in his court to sing for him every day of his life. George II of England also felt a profound sadness that could only be cured by music.

When government and society fell into disarray in ancient China, they tried to reform the degenerate music. Does such "reform" sound familiar?

China

The ancient Chinese believed that music was the basis of everything, that all civilizations are shaped and molded according to the kind of music performed. According to Confucius, if the music changed, then society itself would change accordingly. Because he believed that music had such tremendous power, either for good or for evil, he gave the same importance to music as nations today attach to economic and military issues.

Emperor Shun traveled throughout his kingdom once a year to ensure that all was in order. He did not check the account books or observe the conditions of the people or listen to the petitioners. Nor did he interview the regional officials or authorities. His method for ascertaining the health of his kingdom was to test the pitches of their notes of music. As he listened to the local singers and instrumental musicians, he determined whether what he was hearing was in perfect correspondence with the five tones then in usage. He apparently did the same to check on his central government, believing it was more important to listen to and check the five notes of the musical scale than to rely on expert advice or review the economic indicators.

If the music in the emperor's territories was wistful or romantic, then the people would be wistful or romantic. If it was strong or military, then its natives better beware. If the music remained unchanged, then the society remained stable. If the musical style changed, there would be a change in the people's lives, and if the music in the villages became vulgar and immoral, then the nation would take on this character—unless, of course, something was done to improve the music.

India and Greece

Other civilizations, such as Mesopotamia, India, and Greece, held opinions very close to those of the Chinese. In Hindi, for example, the word for both God and music is *nada.* The Hindu people believe that "sound is all," hence the high value they continue to place on chanting. The emphasis on chanting as a major aspect of prayer runs through virtually every religion. "When we sing, we pray twice," states St. Augustine of Hippo. In Hinduism, the god Shiva Nataraj presents the lure of the cosmic dance as the vibrations of all matter. His stamping feet create the rhythms of creation, of the breath and the heartbeat, of the rotation of galaxies and planets and their suns and moons.

The patterns of Shiva's choreography translate into musical harmonies, the changing of the seasons, and the structure of snowflakes, sunflowers, and crystals. His uplifted arms shake the *damaru,* the small hour-glass-shaped drum that is made by joining the skulls of a man, a woman, and a crown and then covering them with skin to form a two-sided instrument. This drum speaks of the rhythmic interplay of male and female, life and death.

According to Hindu tradition, rhythmic vibration is not simply an element of creation; instead, creation *is* rhythm. Each individual is a manifestation of rhythmic patterning, embodying both collective and uniquely personal rhythms. Every member of our species has the same archetypal pattern.

The ancient Greeks believed that cosmic music accompanied the heavenly spheres in their revolving dances. In the ancient world, the heavenly spheres were the seven "planets" (meaning "wanderers"), which included the sun and the moon as well as the other members of the solar system that were visible to the naked eye—Mercury, Venus, Mars, Jupiter, and Saturn. In Hinduism, the vibration pattern of the deity is believed to exist within the mantra, the yantra (a visual image or geometric design) and the mudra, a sacred gesture. Movement patterns combine in dance, creating geometric patterns, a three-beat rhythm being the simplest. This can be seen as a triangle, four beats as a square, the circle being a continuous movement. The *om* mantra of many cultures supposedly quiets the meditator's mind by using the unbroken circle to achieve a state of oneness with the universe. Many dance traditions tell us that certain geometric designs and rhythms contain the living presence of gods and goddesses, and that these forms of rhythm communicate with the spiritual realm. An ancient Sanskrit hymn has it that "your body is a sacred vibration."

In Haiti, a design, or *vera,* represents each deity. Some rhythms are drummed to summon the presence of that desired deity.

Ancient belief systems might easily be seen as the transmigration of rhythmic patterning to current scientific research such as in wave theories, quantum mechanics, and string theories.

7

Songs of Praise

Speculation about the power of music has engaged many thinkers for centuries. A common theme is that the existence of musical tone precedes material existence. This speculation asserts that the whole cosmos is audible in its superior mode of existence.

In "La musique explique," musical scholar Fabred d'Olivet asserts, "Music is the knowledge of the order of all things, the science of the harmonic relationships of the universe; it rests on immutable principles which nothing can assail." He also believes that music played a vital role in the early history of man, in his worldview and politics. Rudolph Steiner speaks of "the gods' cosmic music of jubilation as the expression of joy of the creation of the world . . . and a lament of mankind's fall."

The Hebrew Kabbalah describes four worlds: the world of emanation or inaudible rhythms; the world of creation or the rhythmical sound or cry; the world of formation or sacrifice; and the world of action, of concrete existence. All four are preceded by silence.

In *Cosmic Music,* edited by Joscelyn Godwin, an essay by Marius Schneider, "The Nature of the Praise Song," says: "Surely one of the greatest errors committed by nineteenth-century religious historians and their successors was the attempt to explain the origin of religious feeling as lying primarily in the human fear of natural forces and to consider supplication as the actual core of the relationship between God and man. . . . The facts show that praise and glorification, the specific act of love in religion, are at least as important as supplication." On the vast hymn literature, it "could be perceived as ultimately designed by poets for the deliberate flattery of the gods." But, he says, "the deception of God by hypocrisy is practically impossible, by its very nature irreligious." He quotes other ancient texts: "First the gods created the song, then the dispenser of sacrifice. Whatever the gods do, they do by song, the song is the sacrifice." He defines the Sanskrit word for singing praise as signified in the Vedic vocabulary as the equivalent of "to achieve, to let something grow through song."

Religious Music

The universality of yearnings for unity and connection with the All has always reposed in some form of spirituality or religion. The roots of religious music go back to the magical signals and symbols in primitive tribal music. For thousands of years before there was written documentation, bards preserved and carried on from generation to generation rituals and laws, knowledge, and legends of heroes and gods. These often became modified along the way as the stories and songs were told and retold, given a new twist here, an addition there, depending on the memory or creativity of the storyteller or singer. Notable exceptions were the Vedas and the *Iliad* which were written. There is a remarkable consistency of style and form in the oral traditions of India, Africa, and even in the West, where our mountain music is one example. Oral history remains widespread, relaying dogma, mysticism, poetry, and music.

For many centuries, song and poetry accompanied planting and harvest. Ritual song forms were thought to assist in bringing forth the world of the spirits with which people communicated through trances, visions, and magic. The original purpose of music was as the language of the gods, and regardless of the musical form, the mystical and supernatural feelings of ecstasy we sometimes get in hearing music—whether it be classical, jazz, rock, or from another culture—is not so different from the feelings of those who prayed to the gods in early times. Ritual psalmody utilized to transmit sacred texts, mystical music for contacts with the supernatural, and music in modern churches all share one basic drive, that is, the search for mystical experiences, for connection with the supernatural, and for ecstasy. This search in modern times takes new forms. Anything goes in contemporary religious gatherings—hymns, *Jesus Christ Superstar,* jazz, rock, symphony, dance. The sacred and the profane have become buddies. Rap and rock music also frequently preach a kind of homily, although it is sometimes expressed in profanity.

The word religion has its roots in connections, and those connections may be realized in many ways, one of which is by way of the past with its traditions and tales, experiences and glories. Experiential and intuitive religion based on the supernatural is at the origin of the entire mystical, animist, Bacchus, and Dionysian traditions. In the animist religions every act of life has a magic and religious property. Most of the big religions started in order to eliminate the supernatural and to restrict religious practice to certain days and times. Even many atheists are not immune to superstitions such as seeing a "bad omen" in, say, a black cat crossing their path.

All of us get a little closer to connection with God, Nature, or the All when we experience music. The better the music, the closer we get to that Supremacy. The urge toward unity is, I believe, a characteristic of our species and perhaps all life. As we shall see, it even pervades molecules, animals, plants, and heavenly bodies.

Hindu and Tibetan tantrism aims to reach the mysteries and unity of the world. This is expressed in certain physical positions and movements as well as in the yogic sublimation and deeper enhancement of sex. The rhythms and pulsations of song and dance enable the practitioner to attain the purity of a high state, erasing the strictures of space and time.

The fundamental Hindu religious song, the *bhajans,* parallels the Roman Bacchanalia, which led to a mystical, manic drunkenness and to the Bengalese *kirtanas,* expressed in chanted hymns and ecstatic dance. Phallic symbols, mixing the sacred and the erotic, were displayed in the Dionysian springtime processions, as they still are in some carnivals. These are a mixture of the sacred and the erotic.

Art music or secular music, as we know it, though existing in a limited sense in ancient times, is a relatively recent (arguably a few centuries at most) phenomenon, taking the spiritual to new and contrasting levels that are quite different from the past.

8

Music As Context

To give a comprehensive and accurate definition of music, one must search for fundamental truths. But as is the fashion in the arts, sciences, and other disciplines, the tendency is to explore minutiae rather than deeper meanings. The wider world of music encompasses history, sociology, religion, technology, politics, and science. All of these are stitched into the connective tissue of music. As they change so does music change, and as music changes it affects all else. By peering beneath the surface of music and transcending the trivia and endless details, one can arrive at the meat of the music, discovering wonders well beyond the temporary pleasures of listening to a work.

What are the influences playing upon composers to make or enable them to write as they do? Why, for example, did Mozart write as he did? Why did Stravinsky compose *The Rite of Spring* instead of a Mozart piano concerto? What were the historical, educational, political, familial, religious, and technological contexts that resulted in these works? How did these and other composers react to these influences? Would Stravinsky have written for such a large orchestra, to take a simple example, if he was not the early beneficiary of a Russian society that heavily sponsored the arts? Mozart had to depend on the simpler instruments available in his time and the largesse of noble patrons. And what was the effect of Mozart having a violinist father who, with other musicians, played music in their home day after day?

What are the traditional, political, and technical bases for electronic rock, rap, hip-hop, gospel, Broadway, etc.?

Do we not become closer to what we assume are our familiar composer friends as we learn of their weaknesses, their strengths, their talents unrevealed even to themselves as they pick up pen or paintbrush?

Is There Meaning in Music and Physics?

Scientists, philosophers, and musicologists have, through the centuries, studied, theorized, and speculated on all aspects of music. The materials, structure, and effect of music have long been known, at least superficially. The meaning of music is more elusive. Though science as a discipline and modern physics are younger than music, the meaning of science and physics has in recent decades made up for lost time. Scientists, pundits, and academic philosophers have been busily raking the subject over the coals of dissension, especially in the wake of quantum theories. But that is a subject for another chapter or two. Regardless of differences, the materials, structure, effect, and meaning in both disciplines cannot be separated from the cultural, economic, and social milieu in which they arise.

Apart from context, there can be no depth of meaning or communication. Written and performed music are bound to the entire gestalt of their period, the ideas that derive from the political and economic climate, the spoken and written language of the time, and the available instruments, materials, and technology. Most of the studies of emotion in music are concerned with mood and the music's association to events of the time. We must consider the composer in the entire context of his life and times, his family background, and even his genes and DNA.

Beethoven was one whose music, directly and by inference, was closely associated with his times, particularly the then current, revolutionary ideas of *liberté, fraternité, égalité.* Less obvious, Baroque music was usually associated with religion and the church. Bach was obliged to write a new work for the church each Sunday. (In his spare time he was apparently busy siring twenty-two children!) What a musical athlete!

The Enlightenment directly and profoundly affected Beethoven. He expressed his personal and musical reaction to Napoleon's predatory war in clear and subjective tones. Less known is that Franz Joseph Haydn was also affected by the same historical period. His *Missa in Tempore Belli* (Mass in Time of War) was composed while Napoleon's armies were pressing eastward.

Though Beethoven had occasional pretensions of grandeur (as in his attempts to claim the "von" of his noble precursors rather than his own "van" Beethoven), he probably would have subscribed to this statement by Haydn: "I have had converse with emperors, kings, and great princes, and have heard many flattering praises from them; but I do not wish to live on a familiar footing with such persons, and I prefer people of my known class." It was only after he left the service of Prince Esterhazy that he entered the Masonic Order (revolutionary at the time) and was no longer required to compose for the liturgical dogma, and was able to musically express his deepest ethical and human convictions. Some of his pieces for chamber groups were so popular that (like Beethoven's) they were played in the streets and beer gardens.

In his *Creation*, Haydn celebrates ethical humanism and the glory of God in nature. It divests God of spirituality and man of the sense of sin. The first part describes the birth of order from chaos; the second, created nature; and the third, human love (Adam and Eve). All this is vividly expressed by both music and word. Unity from diversity is built into his divine prologue. It is through human love that life is redeemed and portrayed as perpetual renewal.

The concept of perpetual renewal seemed born out by Haydn's long life (1732–1809) and enormous output. He was twenty-four when Mozart (1756–1791) was born and composed some of his greatest works and lived for eighteen years after Mozart died. While the child Mozart was already composer and virtuoso performer, traveling all over Europe but never acquiring a secure position, Haydn rarely moved far from Vienna. The one thing they had in common was their humanitarian outlook.

As Alec Harmon and Wilfred Mellers say in their *Man and His Music:* "Haydn saw human beings against their environment, in relation to Nature and a humanistically inclined God. But we know from Mozart's vivid letters that almost as a child, people and music absorbed his whole being. His pen pictures of people are astonishingly acute, neither bitter nor sentimental. We should not be surprised that Mozart was fundamentally an opera composer." I take the view that *all* music tells a story, albeit not explicitly. Listen to Mozart's glorious piano concerti or any of his other exquisite works. Of course the meaning or scenario will vary from listener to listener.

Igor Stravinsky, whose musical approach is the antithesis of that of Mozart and Haydn, once said, "I consider that music is, by its very nature, powerless to express anything at all, whether a feeling, an attitude of mind, a psychological mood, a phenomenon of nature. . . . If, as is nearly always the case, music appears to express something, this is only an illusion, and not a reality." I had a personal experience of his approach to music when I was playing under his direction. I asked him how to play a particular passage in one of his compositions. His reply: "No expression at all, just the notes."

Some modern conductors, Roger Norrington for one, seem to be following Stravinsky's dictum. They slavishly follow the notes and metronome markings. In my view, this clouds or perhaps obliterates the human element, making what could be an emotional performance into a dry, academic performance. Proponents of the literal approach claim,

however, that we should let the music speak for itself, and they religiously follow Beethoven's metronome numbers. So Norrington takes
Beethoven's music at breakneck speed and eschews vibrato. Luckily,
Beethoven survives this mad race. Many string players bitterly denounce
this violation of common violinistic practice. I am tempted to join their
denunciation for yet other reasons: First, Beethoven was deaf when he
introduced these numbers and, second, not being cold and academic
himself he could hardly have intended his music to be so bloodless. Another reason has to do with my experiences of giving first performances
of composers' works. To a man or woman, each was uncertain about precise and exact ways to perform this or that passage, including not only
tempo but even the notes and chords! They usually welcomed a conductor's suggestions. Unfortunately, views parallel to Norrington's take root
in musical cults that reflect the twenty-first-century addiction to technology and antihumanism.

Stravinsky, in trying to get away from the melodic style of Brahms
and Tchaikovsky, began reducing melody to rhythm. But even Stravinsky's music is meaningful, in its Russian, emotional, heritage way. This is
particularly evident in his most popular works, especially in his ballet
music such as *The Firebird, Petrouchka,* and *The Rite of Spring.* These
works, written for choreographers Sergey Diaghilev and George Balan-
chine, as well as some later compositions, were born out of folk themes
and primitive rituals. Anybody who has heard these compositions will
easily sense their very human elements, orgiastic as they are. His middle
and late music was his continued revolt against the musical heritage of
the Enlightenment.

Benjamin Britten was comfortable in his roots, the grand British
choral, hymnal, and poetic traditions. His music is difficult to categorize,
yet much of it marks the realization of what it feels like to live in a world
of faith. His *War Requiem, Peter Grimes, A Boy Was Born,* and *Serenade*
are pure musical and operatic poetry, often tales of innocence turning
into cruelty and tragedy. His was one of the greatest gifts of any twentieth-
century composer. Indeed, one of the most satisfying performances in
my career as a concert artist was performing his *Serenade,* for tenor,

horn, and strings, in New York. A frightfully difficult work for my instrument, I never got it perfectly right at rehearsals, but I was so inspired that in my two public performances at Caramoor I was letter-perfect, sailing through the most difficult passages with ease and daring. I took chances in dynamics, extending from whispers to shattering fortissimos, and these in some of the very highest notes written for the horn. Rarely am I completely satisfied after a performance, no matter the public and critical approval, but in this one even I could not find something I thought I could do better.

After this New York concert, I was invited to London to perform a new Britten composition with pianist Hepzibah Menuhin, a pianist on a par with her magnificent violinist brother, Yehudi. Britten came down from Edinburgh to guide us and at one point sat down at the piano to demonstrate his conception. I swear that he made the piano sound like a symphony orchestra! He was a worthy legatee to Purcell and even Shakespeare. He and his partner, tenor Peter Pears, were aesthetes par excellence.

America was a Johnny-come-lately as far as classical composers were concerned. The first authentic American composer, Charles Ives (1874 1954), was born in Danbury, Connecticut, a town where his father directed the local band, giving Charles training in conventional harmony and counterpoint, with a respect for the "manly" composers such as Beethoven. But Charles was more interested in the native sounds of small-town New England, the four bands that played different marches simultaneously in the town square at celebrations, the players left behind while marching through the streets on July Fourth but continuing to play anyway, camp singers loudly singing different versions of the same hymn, and all this mixed with the sounds of water and wind. He made his rather good living as an insurance salesman, which paid for his real love, writing music that mimicked his environment. Ives's father, when asked how he could stand hearing the local stonemason bellowing off-key at camp meetings, replied: *Old John is a supreme musician. Look into his face and hear the music of the ages. Don't pay too much attention to the sounds. If you do you may miss the music.* Ives has been compared to the poet Walt Whitman, who in his poetry also celebrated the American people.

Among Ives's compositions in which I was involved, I vividly recall two special occasions. One was at Carnegie Hall where Leopold Stokowski and I and another conductor joined in conducting Ives's Fourth Symphony. This work had so much going on at the same time, including different tempi played simultaneously, that Stokowski asked for double assistance. The other event was in Avery Fisher Hall, Lincoln Center, where I produced and conducted a program for Eleanor Roosevelt's one hundreth birthday. I chose Ives's *Unanswered Question* and commissioned Nancy Shear to write a script. (Nancy has always felt close to Mrs. Roosevelt and has been active with the Roosevelt family in Hyde Park, who maintain her legacy.) Since I had worked with Mrs. R on civil rights when she was alive, I too had and have enormous fondness and respect for this great woman. I engaged two actresses, Zoe Caldwell and Geraldine Fitzgerald, both of whom had portrayed Mrs. R on television and in film. They alternated narrating her sayings, which artfully integrated with the music.

From ancient Greek times to the present, philosophers and critics have, with few exceptions, affirmed their belief in the ability of music to evoke emotional responses in listeners. Most of the treatises on musical composition and performance stress the importance of the communication of feeling and emotion. The music of South American composers such as Heitor Villa-Lobos, Carlos Chávez, and Alberto Ginastera is unashamedly emotional, rich and even violent. Most composers demonstrate their faith in the affective power of music by the notes they write and by the expression marks inserted in the score. Some twentieth-century composers, such as Gustav Mahler, have been very explicit in their directions as to how to play the music and what effect they want to achieve.

Listeners, past and present, consistently report that music does arouse feelings in them, though they would have difficulty attaching words or specific associations to describe these feelings. Emotional states are much more subtle and varied than are the crude and standardized words that we use to denote them. Conduct that might to an observer appear to indicate the presence of an emotional response might simply be the result of the subject's daydreams, his observation and imitation of the behavior of others, or his beliefs as to the kind of behavior that is appropriate and expected in a given musical or social situation.

In concerts I have given for small children, when asked what feelings they had from the music or what they thought the music was saying to them, they would give a generalized answer such as "sad" or "happy," or "makes me feel like dancing," or "marching," or "dreaming." Some would pick up on the responses given previously by a more outspoken child. Sometimes they would create a rather interesting story related to their own lives and experiences. Some try to give answers they think are expected of them, and others reflect a general impression or sensation, which arises out of the family and social context. One youngster said that the music reminded him of Khrushchev banging his shoe on the desk at the United Nations!

On the physiological level, music evokes definite responses. It has a marked effect on pulse, respiration, and blood pressure. It delays the onset of muscular fatigue, and studies have shown a clear effect on the psycho galvanic reflex.

A study at the University of Chicago found that the pleasure of music "arises from the artist's playing with forms and conventions ingrained as habits of perception both in the artist and audience. Without such habits there would be no awareness whatever of the artist's fulfillment of and departures from established forms. But the pleasure derived from style is not an intellectual interest in detecting similarities and differences but an immediate aesthetic delight, which arises from the arousal and suspension or fulfillment of expectations which are the products of many previous encounters with works of art."

These conclusions are not so clear-cut, since similar responses go back hundreds or more years in starkly different cultural climates. Can it be then, that the answers may be found in the inherent nature of music itself? My use of the word "inherent" may appear to be a contradiction of previously expressed beliefs, though I am speaking of the cultural superstructure of music only and not the fundamental base of music as waves undulating throughout the universe. Cultures are subject to change, and the music may lag behind, or it may race ahead to predict the future.

The superstructure contains such ephemeral and passing musical phenomena and genres as classical, jazz, popular, rock, gospel, ethnic, new age, and lounge music, to mention but a few of music's many facades. All will eventually enter the musical archives for future research by scholars and the curious. The revival of "old" music, as we have done with Baroque music, manifests nostalgia for a past that we glorify. In a strange way, it is a yearning for a new and better future. The "meaning" of "classical" music, rock and pop lyrics, and musical forms will hardly be relevant even a century from now. Nothing is static—not music, not the universe.

Certain musical phenomena may have a longer shelf life than others. For example, the reactions and expectations that arise out of certain sequences of harmonies, such as subdominant to dominant and back to tonic, seem to be fundamental to all music. Meanwhile, a "resolution," long delayed, like in a mystery story or a film, provides a certain pleasure due to the frustration and anticipation of what will happen next and what the final outcome will be. The best performers make use of this tendency by giving the moment its due while developing a "long line" of continuity overriding all the changes and embellishments.

At concerts, regardless of the musical style—whether it be rock, gospel, or teen-oriented pop, where the audience reacts with a high degree of extroverted enthusiasm, or, on the other hand, meditative "new age" music or the slow, contemplative movement of a Mozart piano concerto, each audience member seems to respond similarly. The meanings may differ from listener to listener, yet there is also a palpable common, if amorphous, bond of meaning for the entire audience in concert hall or stadium.

How About Meaning
for a Composer, Soloist, Conductor?

A composer is like an architect who puts together an entire concept, using whatever materials decided upon (bricks, stone, clay, wood, glass)
and fitting the various spaces these materials enclose into the grander
plan. The performing musician must follow the composer's "blueprint,"
and skillfully knit together individual phrases, sequences, episodes, and
movements into the whole. He must consider the keys, counterpoint,
and harmonic sequences, as well as the all-important themes and
melodies, making it all into a tightly woven presentation. As in literature,
each sentence should bear on the paragraph, which should stay in context of the chapter, which must make sense in connection with the entire
story, article, or book. The very word "symphony" stems from the Greek
sym ("bound together") and *phone* ("sound").

The soloist, more than an orchestral "chair player," must also have an
architectural vision of the music, while the conductor's responsibilities
overarch those of all performers. He must know every single note played
by every member of the orchestra, hear every mistake or note played out
of balance, and know the possibilities of each instrument and each
player's ability. Then, with the ten or the one hundred players under his
baton, he should achieve a result beyond the limit of what each player
and the group think they can achieve, the whole being greater than the
sum of the parts. A great conductor goes beyond even all that and
reaches into the music and into the composer's "soul," to so interpret the
music as to inspire players and public to new heights of understanding of
what it is to be human.

Unlike architecture, though, music moves in the flow of time. The
horizontal flow of time, in the Newtonian sense, is a key element of human experience. A well-thought-out (and felt!) musical concept can result in an experience of inevitability or, in some cases, timelessness. The
opening twelve bars of Beethoven's Ninth Symphony are a splendid ex-

ample of a true artist's use of the passage of time. The open fifths and octaves, the most basic intervals, beginning pianissimo, *sotto voce,* in a few
instruments, are gradually reinforced by additional instruments, and all
make a rapid crescendo in the eleventh bar, reaching a crashing fortissimo in the seventeenth bar, where we finally hear the entire orchestra in
full arpeggio force.

These bars are suspenseful because we as listeners know not what lies
ahead. When I conduct this work, I visualize an entire scenario, the vast
dark spaces of pre–"big bang" time gradually filling up with the materials of the universe, packed so tightly that they explode in a (re)creation
of the universe.

Not long ago the big bang was believed, by the majority of physicists,
to be the beginning of it all. Now many scientists agree that there may
have been something before. Whatever that something or those some-
things turn out to be, there will be something before that. And, on the
other end, after everything that we now know of or can imagine is gone,
there will be something after that. All will change, including music and
physics. All is process and change. Even music. Did Beethoven intuit the
science of the twentieth century?

The opening two fortissimo chords of his Third Symphony, like the
big bang or like buglers announcing the entrance of a king or president,
certainly get your attention. They establish no pattern of motion, so I
conduct them in free time. They forcefully open the curtain, as it were,
of a universe or a highly dramatic opera. For me, the remainder of this
operatic symphony unfolds as a saga chronicled in a long narrative,
telling of the many experiences in life, from ghostly comedy to tragedy to
triumph, from the creation of man to his eventual victory over forces of
repression. Beethoven's Seventh and Ninth symphonies end similarly,
except that in the Ninth the chorus adds explicit words on top of the climactic music. Here is music-meaning writ large!

In such great works, we achieve a compensation and rationalization
for the deadness, the suffering, and the frustration of real life. We enter
a magical, unthreatening world without the fragmentations that isolate
us in a competitive society. On the other side of the coin, trash music

adds to the crippling of human beings, as does much "commercial" music and, sadly, the commodification of "good" music.

Music does not lie. Folk music and some popular music arises out of the real lives of common people. Some classical music and opera sits on the fence. They may have arisen in an honest context but, as they have been taken over by the very rich, they have fallen into a different category. Who else but the very rich can afford the astronomic ticket prices for opera performances! Naturally, there are many interbreeds; rap may be considered one of these. But here again the commercial world gobbles it up as soon as profit possibilities loom. The result is a quick change from a people's manifestation into a commodity. Music has many faces.

9

Music's Dagger

After a couple of centuries of wonderfully gifted composers such as Bach, Beethoven, Brahms, Mozart, Mendelssohn, and Schubert, a countercultural reaction, starting in 1907 by Arnold Schoenberg, stretched the system of musical key centers to its extreme limit. In the 1920s he abandoned keys altogether, developing the serial system of composition, which uses twelve tones equally (supposedly to be democratic and non-hierarchical). Due to the unattractive music that resulted from this system, protests were vehement, and large portions of the audience were alienated from going to concerts. Classical music came to have the reputation of being highbrow and obscure. Popular music came to play a much larger role, especially in youth culture.

Popular music is usually characterized by extremely simple, repetitive rhythms and elementary harmonies—

tonic, dominant, and subdominant (chords based on, respectively, the first, fifth, and fourth tones of the scale). The personalities of certain performers and the words in their songs react against established norms of behavior and serve as a rallying call to countercultural movements. Many pop works are disappointing in live performance unless they are propped up by all kinds of extramusical entertainments.

Before the Victorian period, saloons and pubs were centers for the creation and performance of what were often extemporaneous songs of dissent, satire, complaint, love and rejection, humor, and bawdiness. Yet it can be said that vocal music, specifically the give-and-take of folk singing, engenders a certain moral quality, a community of effort. The competitive concerts in England's monster choral festivals (such as the 3,625 performers who packed into the Crystal Palace in 1862), and the singing competitions that inspired Wagner's *Meistersinger* have continued in a more modern sense to this day. To the extent that these festivals and competitions create the love of and demand for greater music, they serve a positive purpose.

Competition has become fierce today; any opening in almost any orchestra for a flute or horn player, or a conductor, elicits literally hundreds of applications. Many musicians complain that these competitions are, to say the least, highly imperfect criteria and can have a deleterious effect on the "losers." For my musical organizations, I depend a lot on recommendations from musical colleagues, whose judgement I respect.

One day I received a call from London: The polite manager of a teenage keyboard player who was winning major rock awards in England would like to come over and meet me. His name was Keith Emerson, which meant nothing to me, although I was beginning to gain a reputation as a classical-music rebel, involved in too many (according to some of my more straightlaced colleagues) questionable innovations.

Keith and his manager had seen a long article in *American Way* magazine about a multimedia concert/show I had produced in Carnegie Hall. This event was the first such in Carnegie Hall and probably unprecedented anywhere. It included a staged performance of Berlioz's *Symphonie Fantastique* and *Lelio,* the Joshua Light Show, a British choir,

Arthur Mitchell's Dance Theater of Harlem, and Broadway actors directed by Burgess Meredith. Multimedia had arrived in full dress.

The British manager and his client came to visit me. Not more than eighteen years old at the time, Keith told me he was interested in adapting the classics to rock. He played my piano, and I was immediately taken by his talent. I began pulling scores from my cabinet that I thought were "rockable." This first meeting led, in the next few years, to our collaborating in recordings and symphonic rock concerts in Queens Hall, London, and before hundreds of thousands of his fans in outdoor venues. Our collaboration resulted in a rock-style recording of Mussorgsky's *Great Gate of Kiev,* which earned Keith a platinum record and millions of dollars. Not a very good businessman, I got nothing for it, though I did get modest payment for other joint ventures.

So this "over thirty" conductor and the teenage star became friends, swimming the ocean together at Brighton Beach and braving the rapids in Germany. He asked my advice on personnel changes (engaging Greg Lake on vocals and Carl Palmer on drums, the trio that came to be called ELP), whether to marry the lovely young Danish woman he was dating, and on other matters. Our friendship ended when he could not handle his sudden wealth and fame. His new behavior included taunting his wife with his female "groupies" who came to his estate in the British countryside, and employing as guardians the motorcycle group Hells Angels, alleged to have been responsible for at least one murder. Few young rock stars were able to maintain normal decency in a world that valued ostentatious wealth above all.

Keith would come aggressively onstage brandishing a dagger and driving it again and again into his keyboard instrument, all the while playing his music on that same instrument, fighting it as if he were wrestling a tiger. Rock groups have become dependent on all kinds of spectacles— light shows, synthesized sound, prerecorded elements, smoke and fire. Elaborate productions included suggestive dancing and other extramusical attractions.

I had the pleasure of working with Yoko Ono, John Lennon supervising, in a concert at Lincoln Center in the 1970s. Yoko wrote a work called *Apple.* She explained her concept to me so that I could direct the orchestra. While John watched silently with a bemused smile, Yoko was deep into her "composition." For the first portion of her work, she had arranged to have a huge standing fan placed behind the last row of musicians. Shortly after I began conducting the orchestra in a Mozart symphony, she turned on the fan full blast, which, of course, blew the music off the players' music stands, creating growing chaos among the musicians. Soon the music came to a grinding halt. The musicians, professionals all, were at first outraged but quickly became amused and only slightly contemptuous.

The second part of her composition and its climax consisted of prior directions to the players. While I conducted a Beethoven symphony, she instructed the musicians to leave their seats one by one and exchange seats and instruments with another instrumentalist in order to play their music and instruments! The displaced player would follow the same procedure. You can imagine that no violinist was about to entrust his or her valuable instrument to, say, a brassy trombone player!

Chaos ruled.

During all this, Renaissance man Gordon Parks watched from the wings, waiting for me to conduct his composition for piano and orchestra. Parks, an African American, was a much-acclaimed photographer for *Life* magazine when that magazine was the most popular in the country. He was also an accomplished painter, writer, and filmmaker. In his late eighties, he is still busily creating in various disciplines.

Despite the fact that Gordon was totally unschooled in music (he couldn't even read notes), his work turned out to be quite lovely. I recorded it, and the recording was given many subsequent performances.

Here we have different levels of creativity, from Einstein and Beethoven to Yoko!

How to Compose in One Easy Lesson

Some musicians do not need extra paraphernalia. I have asked classical composers how they compose music. Few knew quite how to answer. A couple of them had said that there is music everywhere and they simply have to choose which part of the "everywhere" they want to put on paper. Of the many choices, we simply choose those bits that will work together in some kind of pattern to express what we want to "say."

If we wanted to make ourselves understood to extraterrestrials, our best language should be music or mathematics, as in movies such as *2001: A Space Odyssey* and *Close Encounters of the Third Kind.* The entire cosmos, including our earth, can be described by mathematics, and since man is the study of all possible products (light, behavior, sound, etc.), we ourselves are examples of organized complexity. That definition also describes music, because music, like all in the cosmos, is composed of waves, either electro-magnetic or radio. The same can be said of science and its "composers." Their compositions cannot be completely understood without understanding the math, or the "music" of the human observer (as described in quantum mechanics).

10

Personal Wars

The twentieth century was an exciting period in science, characterized by one important theory after the other: relativity, the uncertainty principle, chaos theory, string theory, and others. Turbulence and innovation in the arts followed a corresponding path; ideas just as far out as those coming from scientists. The sixties and seventies opened the curtain on an explosion of the arts, into the gutter as well as into the peaceful heavens. Allen Ginsberg introduced expletives and street language into poetry, and was at first denounced by academic poets. He made poetry a live craft, giving poetry back to the people. Here there is terra incognita still to be explored; anyone, if they have attended even a single poetry reading, will be disturbingly aware of the gap between performer and listener in this developing but still most intimate and introspective art. A prose revolution arrived with a rush as the great Irish writer James Joyce published his turgid books, incomprehensible to all but the dedicated scholar, who was well rewarded for his efforts.

Rock had its heyday with a bombardment of the senses, a narcotizing of pain, wails and "freaking out" of a lost generation, and the yearning cries of the dispossessed. With the temporary waning of the rock phenomenon, we discovered in these orphans of our culture a new regard for the classics as they looked for some kind of stability and security. These graduates of the psychedelic revolution reached nostalgically back into the twenties, thirties, and forties for their styles of clothing and hair, reviving early Broadway hits, experimenting with Far East religions and philosophies, making pilgrimages into guru lands, and the moving "back to the land."

Talented and creative young rock stars in the early seventies began to mine the treasures buried in the world of the classics. Their astute managers realized there was money to be made here.

A New Ball Game
or Just the Same Old?

Young people found answers in the self-realization movement, with the "Jesus freaks," and in religious sects. The "me generation" arrived in full force. The young and their elders yearned for something or somebody of lasting value to respect and believe in. That need was expressed in much of the new music and art.

Artist revolutionaries became mired in the very filth they saw about them, expressing their frustration and anger by revolting via their lyrics and lifestyles. Visual artists were no less revolutionary, though "anarchistic" might be a better word. One depicts Jesus as a naked woman. Another creates an image of the Holy Mother covered in excrement. Their frustration and feelings of impotence imploded in a world they saw as stolen from under them by an unfeeling and rapacious corporate power structure. A few found a more constructive outlet for their talent and energy. Others surrendered to commercial lucre.

A notable exception was an African-American musician who exemplified an important, ongoing period in American history. Paul Robeson was erudite and sensitive, physically impressive, larger than life, and eminently talented. If he had been born white, it is likely he would have been showered with public acclaim and monetary rewards. Scholar, actor, magnificent singer, football hero, and ardent social activist, Robeson was ahead of his time, an outspoken advocate of many human rights issues (though the term was not widely used then). His dalliance with Soviet Communism was an artifact of the hope, if not the reality.

In a *New York Times* book review of *The Undiscovered Paul Robeson,* by Robeson's son, the reviewer ponders: "It is difficult in the year 2001, knowing what we do about Soviet history, to understand why Paul Robeson and others supported Soviet political actions with such unqualified vigor through the 1930s." This query echoes across the political spectrum. The answer is easy. Negroes, as they were then called, were victims of lynchings and mob "justice" meted out at night by the Ku Klux Klan with the complicity of sheriffs and citizen posses. Entire black communities were terrorized, with venom spewed especially at educated, "uppity" Negroes, especially famous ones such as Robeson. No matter that he was an international star and celebrity in sports and music, with scholarly degrees as well. Trying to attend one of his concerts, many had their cars trashed, escaping only with their lives. License numbers of those trying to attend were taken down, further frightening the potential audience. Their crime? Coming to hear Robeson's magnificent voice.

Is it any wonder that many looked elsewhere for the key to a better, more sane, and more just solution to the problems of society? Young people were vulnerable to the words of utopians and to revolutionary writers such as Karl Marx and Vladimir Lenin. Students and artists joined the poor in a dream that was hard to let go. To be sure, some of us held on to this dream too long, but I can't help wondering if millions of hopeless people today, more than half a century later, might not welcome the vision of a better, more honest, and more secure society, call it whatever ism or party you wish. A dream would be no worse than the widely pervasive cynicism infecting so many people. Music still expresses that dream.

11

Conductors, Consistency,
and Change

"I never use a score when conducting my orchestra.
Does a lion tamer enter a cage with a book on how
to tame a lion?" —DIMITRI MITROPOULOS

Listening

The greatest lesson I ever received, as a musician, performer, conductor, or layperson was how to listen in depth. There are whole layers of listening. Much of these are far deeper than we ever realize, such as listening to every single note in all its breadth, depth, resonance, intonation, timbre, relation to other notes, et cetera. At every rehearsal, before I begin conducting, I first get the eye of every single player in the orchestra. This is, in a sense, an acknowledgment of their importance in the general scheme of things. Moreover, throughout rehearsals and performances, I continue to listen to each player as well as to the orchestral whole in all its dynamics and interrelationships. They know I am listening to them, individually and collectively, and so are encouraged to do their best. In the wondrous magic of working together, we create the music.

It has come back to me more than once that the highly individualistic players of the New York Philharmonic have said, "That Eger has some ears!" They meant that I was able to hear every individual of the orchestra through the sound of the whole. It is not that my ears are any different from others; it is just that I have trained myself to use them, keenly. We take for granted our listening abilities, as we do our other senses, and our raucous environment has desensitized us. "Civilization" has closed off much of our inherent, animal-like sensitivities.

Should a Conductor Be Consistent?

An important issue both for a conductor and for a physicist is how consistent they should be. David Hilbert devoted most of his life trying to find some consistent form of mathematical logic, but he never found an absolute proof of consistency. Bertrand Russell, one of the giants of math, philosophy, and science, gave us a famous epigram: "Mathematics is the subject in which we do not know what we are talking about or whether what we are saying is true." Yet orchestras often request consistency, and most conductors try to achieve consistency in tempo and in rhythm. I contend, however, that the greatest conductors are inconsistent, of course within a frame of reference.

For each portion of music there is a certain tempo that "feels right." One factor is the playability; certain tempi for specific instruments can be played easily and sound best for a section and/or for the whole orchestra. A gradation of faster or slower can make all the difference in the sound and in how well the passage can be played. Take for example the second movement of the Beethoven Ninth Symphony, which goes quite quickly, or the opening of Mendelssohn's Italian Symphony. The tempo must be quite precise within a very limited range; otherwise, it becomes awkward, especially for the strings. By the same token, you can start a movement of a symphony at a tempo that feels right at that point, but a few pages later you might discover that this tempo no longer works at all. Therefore, before you begin, you must decide on a tempo that works for the movement as a whole and then make slight adjustments as you go ("rubato"), according to tradition and your taste.

A fine conductor exercises a certain amount of discretion, and these decisions differentiate a great interpreter from one who is pedestrian. There must be a feeling of cohesion in order to make a short statement, much less a long one over the entire composition. If one has a soloist, one should pretty much follow the soloist and the soloist's interpretation, although some headstrong conductors have been known to vigorously argue the point. A famous incident involved pianist Glenn Gould playing under Leonard Bernstein. Bernstein respected Gould but differed on Gould's unusually slow tempo. They discussed it politely, but Bernstein, as a good host, gave in, albeit reluctantly.

Interpretation in music is also influenced by the period and conditions. There was, for example, the practice of emasculating young boys in order to produce the castrati, popular through the times of Domenico Scarlatti and J. C. Bach. This barbarous practice produced remarkable adult male treble voices but was fortunately ended by humanitarian sentiments arising out of the Enlightenment. Performances of the music written for castrati are frequently given to female singers, though the music is hauntingly beautiful if one is lucky enough to engage the rare male countertenor born with the ability to sing this music.

Being Human

One of the reasons Beethoven was a great composer is that he seems to be more human than most of us. Aside from his musical intelligence and technical expertise, he connects us to the deepest, most hidden corners of our fear, our anger, our hate, our guilt, our love, our gentility. Only one of many examples of his magnificent music is his compassion in the Ninth Symphony as it soars to the very heavens. When the bass baritone declares, *O Freunde, nicht diese Tone! Sondern lasst uns angenehmere anstimmen und freuden vollere.* ("Oh, friends! No longer these sounds of sadness, of misery and tragedy, let us instead tune up better sounds, to a world of joy.") This theme runs throughout as he musically emphasizes the "all" (offbeat) in "*all* humankind." His own yearnings come through when he says, "Surely there must be a loving father in the heavens." Clearly, Beethoven had his own problems with his father.

A conductor doesn't have time ("time is money"!) to discuss all this with the orchestra. It is his job to be the organizing principle. Orchestras do not want to hear rationalizations and lots of talk. They want communication on a much higher level, the music. The conductor must speak with his body, his baton, the fire in his eyes, and the purity in his heart, producing waves and emanations as understandable to the orchestra as mathematics is to physicists. There are almost as many interpretations of the Ninth as there are conductors.

One can find absolute consistency only within a particular system or framework. This principle is also true in some music. In conducting an opera such as *Carmen,* there are certain traditions of *fermati* (holding a note), *accelerandi* (hastening), *ritardandi* (slowing down), and the like, which are largely observed because operatic singers are accustomed to that tradition. The conducting motions from the conductor's position in the pit must be observable to the singers and choruses onstage as well as to the extreme ends of the widespread orchestra in the pit. Sometimes, as in a Sousa march, one can get the tempo started then barely conduct and let the band or orchestra carry the momentum. Some music requires tiny beats from the conductor in order to achieve extreme precision from the orchestra.

The conductor must defeat Newtonian time by simultaneously living each instant in the past, present, and future of the phrase and throughout the work. He must at all times be aware of the overall shape of each phrase and movement within the composition as a whole. The conductor must maintain the overall vision of the music. The musicians understand the deeper meaning when I tell them that the goal is simply to "make music." As to what that consists of, well, you will recognize it when you hear it! There is a certain ineffable human communication that should come through in a performance.

A famous New York Philharmonic conductor had a fabulous musical memory—he conducted without a score—also had an excellent baton technique. With all that, he still bored the musicians, who felt that he was not really digging into the heart and soul of the music. I rarely attended his concerts. On the other hand, I have played under, seen, and heard conductors with poor baton techniques who managed to make very good music. For example, there's a very old recording of Beethoven's Ninth Symphony led by German conductor Wilhelm Furtwängler. The recording is not up to par, the orchestra is not quite together, there is an occasional wrong note, and it is not exactly in tune. However, it is one of the most marvelous renderings of the Ninth. There is common agreement that this conductor really made music. Overall, there was something profound, spiritual, and human in his interpretations.

Leaders of the Band

"I am become death, the destroyer of worlds." —BHAGAVAD GITA

During the development of the atomic bomb, J. Robert Oppenheimer was chosen to head the Manhattan Project, not only because he was a brilliant physicist and mathematician but also because he was equally adept at organizing and perceiving the nuances of the team he led. As much poet as scientist, Oppenheimer had to "balance" his forces when he was "conducting" the Los Alamos experiments, and, like a great conductor, he quickly became acquainted with the psychological makeup of each of the eighty to ninety members of his "orchestra." This included a hands-on approach whereby he reached out to the technicians and machinists who manufactured parts for the project, not just to the theoretical team, giving every individual a sense that he was a significant part of the project.

Oppenheimer orchestrated the team with extraordinary skill and insight. It was as though he were making music. The balancing of players and instruments is a huge part of conducting. In many respects, greatness in any leader requires sensitivity to all aspects of the environment, human and otherwise. No matter how talented or skilled one might be, constant attention must be paid to all the players. Some musicians or scientists must be massaged and coddled, while others respond to restrained, almost severe, treatment.

Edward Teller, of superbomb fame, was a particularly difficult member of the team and had left it when, in the midst of playing a Mozart sonata, he heard the news of the first successful test of an atom bomb that used nuclear fusion (instead of fission, as the first ones had). He had lobbied for this alternate path to a hydrogen bomb.

The managing of the orchestra starts before one walks to the podium. Choosing and assigning a hierarchy of musicians requires insightful evaluation. Good section leaders can make your task easier. Players and sections must be worked and integrated. Some players or sections respond to compliments, others do not. The conductor may acknowledge with a word or a nod, being careful not to over-compliment. Everybody likes to be acknowledged, but it becomes a balancing act, where talent, ego, and emotion all come into play. The "CEO" of the orchestra must be sensitive to the daily mood changes of the orchestra. Is there relevant gossip that might affect performance? Did this or that player have an argument with his or her spouse? Was a player bilious? (Most performers play better on an empty stomach.)

The physical environment is very important. Each hall is unique, depending on different acoustics, size, construction and materials, temperature, humidity, air-conditioning, etc. Each audience is also unique, and audience demographics are crucial factors. Halls in the Baroque period were much smaller (concerts and recitals were even held in drawing rooms), accommodating the weaker instruments and smaller orchestras of the time. Gregorian chant sounds best in great cathedrals. Brass bands are meant for the outdoors. All these nuances are the stuff of which music, audience, and life are composed. As in physics, balance is a constant factor. One of the major differences between a great conductor and a good conductor is the degree of attention to the balancing of the various "voices"—inner voices (the harmonies) as well as the melodic voice—of the composition and the instruments in such a manner that the essentials of the music emerge—the harmonies, chords, counterpoint, subthemes, et cetera.

The personal lives and group dynamics of the orchestra members also affect the music. What does the player or group of players bring from the outside world that can be funneled into the music? The conductor senses and listens. When a young player tells me that he or she is nervous, I say, "Good, it is a human quality; put it into the music." If a tragedy or a joyous occasion occurs, I encourage them to put those emotions into the music. All is grist for the mill of music, for music is big enough to encompass all emotion.

Otto Klemperer and Technique

When I was a young French horn player in the Los Angeles Philharmonic, we had an excellent permanent conductor, Alfred Wallenstein, who was also a good musician (these are not always synonymous). He was versatile and technically proficient. His baton technique could clearly communicate his notions about the music. He had been first cellist under Toscanini, no mean accomplishment.

One day Otto Klemperer was our guest conductor. Klemperer was not well. He had suffered a stroke and had been mugged in a nightclub. The right side of his body was so paralyzed that spittle would seep uncontrollably from his mouth. Worse, his "good" arm lay at his side, useless. His left arm shook uncontrollably. But when he conducted there was fire in his eyes, and it motivated us in a singular way. Over all these handicaps, his intensity and love of the music inspired us to perform better than we did with our regular conductor, who was graced with two functional arms but just did not have the stuff of greatness. Klemperer reached inside us. We somehow played together despite lack of a clear beat. I suspect he reached inside the audience, too. This is not to demean baton technique or technology in general, I only wish to point out the value of combining technology with human values.

A Newtonian approach would work quite well for dance—ballet and modern—since the chief requirement of this art form is consistency. Once a tempo is set, the dancers rehearse to it daily, and the conductor must keep the same tempo. He or she must also be very sensitive, to "feel" the movements of the dancers, entering their bodies, so to speak, and dancing with them. Unity must exist between the musicians, dancers, and conductor. A conductor can actually damage a dancer physically by unexpectedly changing tempo and causing her or him to strain or stumble. The music must be exactly the same, performance after performance. I find that conducting for dancers is a challenge for the first two or three rehearsals, but soon after, it becomes a dreadful bore, unless one views it holistically and holographically, adding the reactions of each new audience (the main factor of change) to the mix. This makes an ineffable difference.

Serge Koussevitzky

Russian-American conductor Serge Koussevitzky (1874–1951) conducted the Boston Symphony Orchestra for a quarter of a century and is credited with making it one of the great orchestras in the world, a distinction it still holds. He also created the Music Center in Tanglewood, Massachusetts, where the Boston Symphony was in residence, performing and teaching every summer. I was a student there, studying under the principal French horn player of the BSO and playing in the student orchestra. Koussevitzky contributed much to the onward march of music, teaching and encouraging young composers such as Aaron Copland and conductors such as Leonard Bernstein, whom he took under his wing. He introduced much new Russian music to the United States; he is credited with giving first performances of music by Shostakovich, Prokofiev, and other Russian composers. Though a profound and insightful musician, Koussevitzky was a poor technician. He had no upbeat and a terrible downbeat, slowly lowering his baton from its raised position instead of giving a sharp downbeat like most conductors. The upbeat should signal the tempo. A downbeat, of course, is crucial because it must clearly indicate precisely where the players should begin, hopefully together. String players begin a note with the bow and wind players with the breath and tongue. A poor downbeat usually leads to sloppy ensemble. The funny thing is, however, that the Boston Symphony was noted for its superb, unmatched ensemble.

How did this happen? How did they know when to begin? We asked our first-chair teachers in the BSO. The whimsical response: "When his baton passes the third button on his vest." This is an inside joke among musicians.

Leopold Stokowski

As we've seen, the conductor must have many tools at his command. He must know the music backward and forward, he must know the characteristics of every instrument and its part in the music being played. He must mix discipline with occasional humor, knowing when to explain something and when to let his gestures, his eyes, and his mien convey the music and his concept. He is an ensemble- and balance-keeper. Above all, he must be a keen listener.

One time during my four years as Stokowski's associate conductor with the American Symphony, we were in Carnegie Hall during a rehearsal intermission. Stokowski was sitting on his high stool at the podium. His eyes were directed at the score on his music stand. My guess is that he was deciding just how he would handle the second half of the rehearsal. He was in his eighties and by this time had two artificial hips. Some of the musicians were still onstage, and apparently an argument was going on between two groups of musicians. It was getting to be rather a ruckus. Stokowski kept his eyes on the music. The two groups of musicians approached the podium and stood at either side of the conductor. Stokowski never lifted his eyes, though it was clear he was listening intently.

One group vociferously proclaimed their side of the argument. The other group jumped in and equally angrily yelled out their point of view. The first followed, a little less loudly, explaining their position, and was soon answered by the second group, also in more moderate terms. Then the first group in respectfully conversational tones further set out their position, followed in the same tones by the second. All the while, Stokowski's eyes were still on the score while he was obviously listening to all that was going on. Both groups, seemingly satisfied, gradually meandered away to their seats or backstage. Stokowski had never said a word, never took sides, but satisfied everybody simply by listening. He was, in addition to his musical skills, a keen psychologist.

This was a profound lesson for me. I had already come to understand, in my conducting career, the great power of listening and had gained the habit of engaging the eyes of every individual member of the orchestra before I would start conducting, whether in a rehearsal or a concert. Everybody needs to be acknowledged, to know that their voice is heard. Once heard, of course, this is not always sufficient. There is the hypocritical "listening" of a politician who could not care less for the opinions or needs of the person, unless they are a contributor or their voice is strong enough to threaten his power and position. In the case of our orchestra, however, the musicians presumably were satisfied simply to be recognized by the supreme judge—the conductor.

Another example occurred when Yehudi Menuhin, a great violinist and conductor, was the guest conductor. The orchestra was onstage, already tuned up, waiting for the conductor to appear. Yehudi entered from the side of the stage and walked toward the podium with enthusiastic applause from audience and orchestra. Stokowski and I were sitting in our usual front box for such occasions. I turned to Stoki (as he was affectionately known) and said, "Maestro, I still think of Yehudi as a wunderkind." Menuhin had achieved his worldwide fame as a child genius. At this time, Menuhin was probably in his mid-fifties, but it occurred to me that it was as a child genius that many people remembered him. Stoki turned back to me and replied, "Maestro [we always maestro-ed each other!], all men are boys." I thought for a moment, then asked: "How about women?" His response: "No, women grow up!"

Many factors go into the respect or lack of same, veneration or contempt, in which a conductor is held by the orchestra. There are certain givens, such as knowing the score and all of the instruments. But when it comes to a great orchestra, the criteria are much more subtle and complex. The New York Philharmonic, an orchestra that I know rather intimately, is a conductor's instrument. Critic Bernard Holland describes it well: "The New York Philharmonic . . . is a 'blank check' orchestra. It can be round and soulful for Sir Colin Davis or Riccardo Muti and uproariously vulgar for Mr. Mehta. The Philharmonic players wriggled with discomfort during the Kurt Masur years which had the characteristics of a boot camp; independent-minded, occasionally arrogant musical soldiers being whipped into shape by a tough German drill sergeant. After years of Zubin Mehta's superficialities, the musicians longed for a master conductor who would lead them into depths not often explored by Mehta."

12

The Audience Revolution

*"If the day ever comes when people no longer talk
about melody and harmony, about German and
Italian schools, about the past and present of music,
then the kingdom of art will probably begin."*
—GIUSEPPE VERDI

"Ultimately I want to integrate everything."
—KARLHEINZ STOCKHAUSEN

UNESCO (United Nations Education, Science,
and Culture Organization) commissioned me to
write an article for the first issue of their new publication, *Cultures*. Volume 1, number 1, was titled "Music
and Society." Other articles in that debut issue were by
Yehudi Menuhin, Luciano Berio, Pierre Boulez, Ravi
Shankar, and Andrew Lloyd Webber. The editor's introduction and the gist of my piece follow:

Music in the late twentieth century became a universal
language, transcending national and regional vernacu-
lars. Musicians everywhere have accepted practices in
which various cultural streams come together, composers utilizing freely the forms that suit best their inspiration regardless of the origin of these forms. European
music is impregnated with Oriental modes and African
rhythms; Eastern music shows influences of Western

forms. New instruments have been incorporated into the orchestra, electricity and electronics have been called on to modify the nature and quality of sound, new scales and tonal relationships have been found. The universality of music is increasingly apparent not only in serious music, but also in popular music; young Americans dance to the same tunes that are heard in Buenos Aires, Budapest or Tokyo.

"A world of music is in the making in which the great wealth of musical languages that evolved throughout history converges to create artistic forms, which convey the variety of feelings, concepts and aspirations of man in contemporary society.

"As the character of society changes, so does musical expression. Traditional societies utilized music one way; industrialized societies in another. Music, by its very nature, is an indicator of the cultural climate and the multiple social factors of the time" (G. S. Metraux, *Cultures* editor).

Music's Uncertainty Principle

In the first half of the twentieth century, the problems of music seemed simpler. A composer could choose to write either tonal music or some variation of atonal music, such as twelve tone, aleatoric, electronic, etc. Twelve-tone music derived from Schoenberg's theory that all 12 tones of the scale should be equal. A laudably democratic principle but an oversimplification of democracy's meaning, and it didn't catch on, at least with audiences. Nor did aleatoric music—anything goes music—as in the Yoko Ono composition previously mentioned. Still, these peregrinations persisted, at least with composers.

Stravinsky skillfully maintained a tonal base until he had to succumb to Arnold Schoenberg's twelve-tone musical revolution. Tonal music became déclassé among the cognoscenti, who disdained it as romantic music. Such self-defined sophisticates contributed to a half century of lost audiences who hated "modern music."

The urge to escape the past and "modernize" the visual arts went hand in hand with that of music. The impressionist movement in art had its siblings in the music of Debussy and other impressionists. After going to a concert of Schoenberg's music, Wassily Kandinsky was inspired to rush home and paint the music, not so far-fetched since Schoenberg also painted. They became fast friends. In dance, Merce Cunningham reacted to the innovations and style of John Cage. Sergey Eisenstein made the leap in film with the help of one of the musical giants of the twentieth century, Sergei Prokofiev.

Not all these innovations were kindly met. Furious audiences walked out of performances of Schoenberg's and his disciple's twelve-tone music despite praise from newspaper critics and adulation from his composer peers, many of whom aped his style. Musicologists had new material for their Ph.D. dissertations, and critics expounded learnedly.

In all this time the public stayed away in droves. Apparently they intuitively agreed with Pythagoras and his buddies about the nature of music, basically constructed out of the harmonic series. This series is omnipresent wherever there is sound extending, as far as we know, throughout the universe. Every tone, high or low, whether emanating from an organ's pipe, the throat of a baritone singer or that of a bird, the tinkle of cocktail glasses when struck together or the clap of thunder, all have their own, predictable series of overtones or sub tones based on strictly predictable arithmetical fractions of frequencies, half, two-thirds, three-fourths, etc. In musical intervals these become octave, fifth, octave, third, fifth, seventh, octave, etc. Later chapters translate these into physics.

When I was a French horn player, some colleagues and I had fun playing "horn" quartets on the long (as much as six feet or more), rubbery California coast seaweed. They somewhat resemble animal horns. We would cut off the small ends so as to achieve four different lengths. As in a trumpet or other brass instrument, the length determines the "key" or pitch, adjusted by the valves. Though the notes on each seaweed were restricted to notes in the series of that particular seaweed, we were able to tune each (by judicious cutting) and made some rather acceptable harmonies by the four of us together blowing our homemade instruments.

Benjamin Franklin created his "Harmonium" by filling glasses with different levels of water, then striking these with a stick or sticks to achieve melody or harmony.

A New Drama

In the postwar period it was small wonder that the arts had become disoriented. The music was mirroring unprecedented changes. Man for the first time in history had the power to destroy his entire species. Slumbering continents, Africa foremost, burst forth with nations emerging one after the other, swelling the ranks of the United Nations. The moon and the atom shed their mystery. Wars had hastened the advance of technology for better or for worse.

Never before had reality surpassed man's wildest imaginings. Where was the theater that could match the drama of Vietnamese children racing down the road in terror while removing their burning clothing; the Israeli mother mourning the death of her child caught in cross fire? Or the dead Palestinian child in his father's grief-stricken arms, as both cringed in the street to avoid "collateral damage" from tanks and guns? Where was the fantastic symphony that could capture the sounds in space made by our astronauts? Where was the painter who could match the graffiti murals that covered the subway system of New York City?

In the sixties and seventies thousands of young "vandals" had collectively created contemporary cave painting, which was a far more vigorous commentary on their time than any isolated artwork. Their powerful statements in swirls of color on moving trains and subway walls seemed to declare: "I am! I exist! In spite of this noisy, dark, dangerous, dirty environment, I will illumine my world with laughter, with life, with color."

Icons of Musical Revolution
and Co-optation

The Beatles, Bob Dylan, and others spoke a refreshing new language. This period was characterized by a schism between parents and child, when "you could not trust anyone over thirty" (my greatest accolade at the time was a headline in *Rolling Stone* magazine calling me "The Conductor You Can Trust"). Young people, as is their wont, were the first to respond, and soon the new language began to reach across generations, temporarily uniting children with their elders. Who could resist the Beatles and some of the rock world's better groups?

Getting into the mood of the time, some friends and I created an organization called New York Aware, dedicated to the idea that the city is theater . . . everyone a player on a common stage, that each block and neighborhood was interdependent with every other "hood." For example, one block could clean up its street only to have the wind blow debris right back from the next block. We were all in this thing together. We produced a citywide Aware Festival, simultaneously in New York's all five boroughs. Tens of thousands of people joined hands in a huge circle in Central Park and sang with Pete Seeger leading. Other musicians performed on several stages. Althea Gibson, a tennis luminary at the time, gave free tennis lessons.

One outcome was an invitation I received from Dr. Andrew Ferber, author of *The Book on Family Therapy* and head of family therapy at Albert Einstein College, to address his fellow psychiatrists. Andy recognized the power of the arts to influence family relationships.

Another consequence of this exciting and creative time (my "rock period"!) was Crossover, a group of twelve musicians made up of string quartet, brass quartet and rock quartet, each genre borrowing from the other. One of our performances was given at Fillmore East, the center of rock concerts, where the best groups performed. The foremost classical impresario of the day, Sol Hurok, presented us at this Vatican City of the rock world in New York's East Village as the opening act for the Grate-

ful Dead. For our unique instrumental combination we made arrangements that I liberated from Bach, Berlioz, Beethoven, and the Beatles'
"Eleanor Rigby." Mea culpa! Today's Grateful Dead might well be the
band Phish, with their tendency for extended improvisation and eclec-
ticism.

If Fillmore East was the Vatican City of the rock world, then The
Dead was the Pope. My niece, Jaimila, was one of a huge fan club nick-
named "Deadheads" who followed the group from city to city to hear
their concerts. Our Crossover paled in comparison, though I did have
the joy of seeing my niece, who traveled from Oakland, California. My
stock with Jaimila soared and my reputation among young people was
made, though my classical colleagues looked askance at my deviation
from high art.

One summer when I was conducting a youth orchestra in Tanglewood, Massachusetts, I heard Carly Simon, in a nearby tent, teaching her
new song to the young campers. She called it "Secret Saucy Thoughts of
Suzy." When I returned to New York, where I was conducting the American Symphony as Stokowski's Associate Conductor, I arranged Carly's
song and juxtaposed it with the Brahms Fourth Symphony for a series of
twelve concerts in Carnegie Hall! At the same concerts for New York
City high school students, I presented a then unknown rock group, the
Elephants Memory, later picked up by John Lennon and Yoko Ono as
their recording group. The mixture of symphony music and rock music
made a great hit.

Prior to our concerts, Board of Education and Carnegie Hall officials
were fearful about security. Teenage crime was very prevalent in those
days, and the prim Carnegie Hall was especially worried since many of
the youngsters were from "tough neighborhoods." Though they planned
on having police in attendance, I adamantly refused, and in all twelve
concerts we had not the slightest bit of trouble. Music had bonded the
varied audiences and "soothed the [alleged] savage breast."

Music crossed over age differences as well. My mixed concerts with
orchestra and groups such as the Rascals, the New York Rock Ensemble,
and others drew audiences of young people *and* their parents, who prob-

ably had never attended a rock concert. The generations were drawn closer together as the adults experienced their children's music and the young people (the first time they came to a symphony concert) realized that they liked classical music. Subsequently, any time teenagers learned of my checkered musical background (working with famous rock groups), I was quickly adopted, albeit with a sprinkling of awe and respect for my ancient age of thirty-odd years.

As with today's Eminem, *8 Mile,* and *Jackass'* television/film show, the commercial world saw a market for angry young talent, opened its huge maw, and gobbled it up to be used in commercials, films, and to advertise corporations and their products, from dish powder to diamonds. The best talent of the time, twentieth-century Mozarts and Beethovens, were drawn to the money to be made in commercial music. Some of their work was and remains stunningly innovative; frequently well beyond the work of many academically "serious" composers.

After a decade or two, new energies emerged, in a multitude of genres. Always in music's vanguard, African-Americans created new forms and new language. The music of young white, black, and Hispanic talent in rock, rap, jazz, country, ethnic, and a slew of other forms reverberated among masses of youth, usually out of their alienation from what many saw as the wholesale hypocrisy and corruption of political leaders, the confused timidity of their parents, and the violence, brutality, and racism in the world.

"Serious" Music

Serious music in the sixties and seventies took a backseat, never to regain the front. As chairman of the Program Committee of the now defunct National Association of American Conductors and Composers (NAACC), it was my end task to choose contemporary American works to be performed in New York City. We perused hundreds of contemporary scores, selecting only a small percentage for our concerts. Judging was an awesome responsibility. These composers had invested their life's blood in their compositions, most of which had to be rejected. Aside from those that were derivative or poorly formulated, many of the scores were made up of geometric designs and squiggles that the conductor and orchestra had to somehow interpret. It was called "chance music." "Anything goes" was the motto in music.

Some more talented experimenters such as John Cage, Ralph Shapey, and more recently, the very gifted Steve Reich, Shulamit Ran, Robert Beaser, and a handful of others stimulated new thinking, opening a crack here and there with some interesting compositions. Frederic Rzewski's newest long composition, *The Road,* is made up of eight sections or "miles" and a novel, music to be "read" or played at home, with the "interpreter" free to react to the "music" as he chooses. The idea of writing a set of variations for piano for domestic improvisation implies a criticism of classical music's formalism. His *The People United Will Never Be Defeated* speaks for an entire genre of protest music.

New York Times critic Ben Ratliff, writing about another contemporary composer, reports: "Its unbearable text dealt with string theory, physics, gravity and vibrations!" He might have reacted differently to two works by composer Peter Jona Korn that we performed during the sixties at New York's Town Hall. Peter had been either ignored or treated with contempt by fellow "serious" composers for writing "romantic music." It was partly my fault because when he offered to write some works for me, he knew that I shared his concern that music be audience friendly.

What About the Audience?

Peter had the last word as he became widely accepted in Europe and became president of a large music conservatory in Germany.

A parallel controversy over uncertainty and chaos theories was shaking the science world, though it too needed public approval. How else to obtain grants? The "user friendly" computer was the most sought after, and the greatest of scientists were very much aware of *their* audience. Many were musicians themselves.

I've always respected the audience, and I usually like the music they also respond to. This excluded almost all of so-called avant-garde music. Music does not need to be simplistic or saccharine to be accepted by audiences—as witness Bach, Bartók, Beethoven, and other great composers. Leopold Stokowski, who introduced many of the twentieth-century works that are classics today, would play through a new composition twice in one concert for his Philadelphia Orchestra audience. Koussevitzky did much the same with the Boston Symphony. Many of these works then became popular staples of the concert hall. But when a persistent dichotomy appears between audience and composition, one must question the quality or timeliness of the composition; for though artists are teachers as well as entertainers, revealing bits of the world in new ways, that world must have at least a modicum of the audience's familiar world. T'aint easy. That's what separates the academic from the great artist. Beethoven's music, for example, was the "pop" music of his time. Some of his smaller works were even played in beer halls. Were his "pop" music performances a precursor to rock concerts? The questions of the sixties and seventies have morphed into the death throes of one symphony orchestra after another while the arts in education is being replaced by sports and television. Yet some heroes remain alive.

13

Beethoven, My Hero

"Socrates and Jesus were my masters."
—LUDWIG VAN BEETHOVEN

What Beethoven and Einstein have in common is the scope of their vision. Both are products of their historical and cultural milieu, and both yearned for a better, more unified world. Both encompassed the universe in their work, though Einstein's vision was of the physical world and Beethoven's was of the spirit— the yearning to be free and to bring all of mankind with him. The vision in his music is patently universal.

Though Beethoven probably does better at the box office than Einstein, I suspect that there is more certainty about the scientist's life than about that of the musician. However, I would wager that Beethoven's music—though easier to listen to than Einstein's equations are to comprehend—remains as much a mystery to the average person as do Einstein's relativity theories.

The private lives of both men have been probed,

prodded, weighed, and examined under a microscope. In ensuing chapters we shall learn about Einstein's theories and Einstein the man. Luckily, many of us remember him in person. With Beethoven, we are fortunate that he left a rich legacy of revealing letters, notebooks, a personal diary, and sketchbooks of his compositions. Beethoven scholars disagree about the interpretation of these artifacts, but those who love his music could not care less about these scholarly musings.

Einstein's life was, perhaps, more a life of the mind, with domination by his left brain, the portion that emphasizes reason and math. On the other hand, Beethoven's brain was right-dominated, steeped in emotion and the imagination, with enough of the left to create his musical "equations." Neither wholly restricted himself to the rational or the artistic, finding a rich balance and unity between the two. Both great men had all sides of their brains working quite well together, thank you. Einstein's love of music and his view of science as an art more than a science attest to a mind plentiful with warmth and imagination. By the same token, Beethoven could never have written his huge output of monumental compositions constructed of billions of tiny notes, ingeniously strung together like the atoms and molecules in DNA, without a strong left brain and a structured, Newtonian mind to go along with his artistic right brain.

Neither man sprang forth full-grown from Zeus's brow. Einstein always paid specific and honest respect to his forebears. Beethoven was more circumspect, and in some instances evasive, though he was fulsome in his praise of Haydn in particular. Though scholars differ in many details of his life, not in contention are the facts that Beethoven's grandfather was the *Kapellmeister* and bass singer at the electoral court and that his father was a court tenor and music teacher.

Beethoven the Man and Revolutionary

Ludwig van Beethoven was baptized on December 17, 1770, though he claimed that he was born in 1772, implying at various times that he was three to five years younger than his actual age. Beethoven's father may also have falsified his son's age in order to better promote his son as a wunderkind. In his later years, Beethoven resisted denying the rumor that he was the son of the late king of Prussia. Evidently, his need to identify himself with royalty was only one of the many complex contradictions both in himself and in the view of others.

Though he left voluminous records of his life and works in the form of letters, scores, and sketches, the hundreds of books and papers written about him differ widely in their interpretation—and even the facts—of his life. As George R. Marek said in his *Beethoven: Biography of a Genius,* "Can one describe music?" I would add: Can one describe a "genius"? This difficulty hasn't stopped authors from heavily romanticizing, speculating, analyzing, and playing fast and loose with the established facts about Beethoven's life.

An exceptional book on the subject is *Beethoven,* by Maynard Solomon. Solomon analyzes Beethoven and his music in depth, mostly with a psychological, Freudian slant. The author was also very well versed in music, having established a record company that released high-quality classical music.

The main "facts" of Beethoven's life, however, may best be revealed in his letters and, of course, in his music. I shall go into some detail about his music in a following chapter, but first let us examine his letters. One letter in particular reveals the musician who is going deaf, the worst fate for a musician. His "Heiligenstadt Testament" was written at age thirty when he was forced to admit to himself that he was going deaf. What a fate for a musician! He wrote this famous letter to his brothers, wherein he seriously considered taking his life. I submit it in its full length:

> FOR MY BROTHERS CARL AND _____ BEETHOVEN*
> Oh you men who think that I am malevolent, stubborn, or misanthropic, how greatly do you wrong me. You do not know the secret cause which makes me seem that way to you. From childhood

*Empty space in the original refers to the brother whose name is never mentioned by Beethoven.

on, my heart and soul have been full of the tender feeling of good will, and I was ever inclined to accomplish great things. But, think that for six years now I have been greatly afflicted, made worse by senseless physicians, from year to year deceived with hopes of improvement, finally compelled to face the prospect of a lasting malady (whose cure will take years or, perhaps, be impossible). Though born with a fiery, active temperament, even susceptible to the diversions of society, I was soon compelled to withdraw myself, to live life alone. If at times I tried to forget all this, oh how harshly was I flung back by the doubly sad experience of my bad hearing. Yet it was impossible for me to say to people, "Speak louder, shout, for I am deaf." Ah, how could I possibly admit an infirmity in the one sense which ought to be more perfect in me than in others, a sense which I once possessed in the highest perfection, perfection such as few in my profession enjoy or even enjoyed. —Oh I cannot do it; therefore forgive me when you see me draw back when I would have gladly mingled with you. My misfortune is doubly painful to me because I am bound to be misunderstood; for me there can be no relaxation with my fellow men, there can be no refined conversations, no mutual exchange of ideas. I must live almost alone, like one who has been banished; I can mix with society only as much as true necessity demands. If I approach near to people, a hot terror seizes upon me, and I fear being exposed to the danger that my condition might be noticed. Thus it has been during the last six months which I have spent in the country. By ordering me to spare my hearing as much as possible, my own intelligent doctor almost fell in with my own present frame of mind, though sometimes I ran counter to it by yielding to my desire for companionship. But what a humiliation for me when someone standing next to me heard a flute in the distance and I heard nothing, or someone heard a shepherd singing and again I heard nothing. Such incidents drove me almost to despair; a little more of that and I would have ended my life—it was only my art that held me back. Ah, it seemed to me impossible to leave the world until I had brought forth all that I felt was within me. So I endured this wretched existence—truly wretched for so susceptible a body, which can be thrown by a sudden change from the best condition to the very worst. —Patience, they say, is what I must now choose for my guide, and I have done so—I hope my determination will remain firm until it pleases the inexorable Parcae (Fates, ed.) to break the thread. Perhaps I shall get better, perhaps not; I am ready. —Forced to become a philosopher already in my twenty-eighth year, —oh it is not easy, and for the artist, much more difficult than for anyone else. —Divine One, thou seest my inmost soul; thou knowest that therein dwells the love of mankind and the desire to do good. —Oh fellow men, when at some point you read this, consider then that you have done me an injustice; someone who has had misfortune may console himself to find a similar case to his, who despite all the limitations of Nature nevertheless did everything within his powers to become accepted among worthy artists and men. —You, my brothers Carl and _____, as soon as I am dead, if Dr. Schmidt is still alive, ask him in my name to describe my malady, and attach this written document to his account of my illness so that so far as is possible at least the world will become reconciled to me after my death. —At the same time, I declare you two to be the heirs to my small fortune (if so it can be called); divide it fairly; bear with me and help each other. What injury you have done me you know was long ago forgiven. To you, brother Carl, I give special thanks for the attachment you have shown me of late. It is my wish that you have a better and freer life than I have had. Recommend virtue to your children; it alone, not money, can make them happy. I speak from experience; this was what upheld me in time of misery. Thanks to it and to my art, I did not end my life by suicide—Farewell and love each other—I thank all my friends, particularly Prince Lichnowsky and Professor Schmidt— I would like the instruments from Prince L. to be preserved by one of you, but not be the cause of strife between you, and as soon as they can serve you a better service, then sell them. How happy I shall be if I can still be helpful to you in my grave—so be it. —With joy I hasten to meet death. —If it comes before I have had the chance to develop all my artistic capacities, it will be coming too soon despite my harsh fate, and I should probably wish it later—yet even so I should be happy, for would it not free me from a state of endless suffering? —Come when thou wilt, I shall meet thee bravely. Farewell and do not wholly forget me when I am dead; I deserve this from you, for during my lifetime I was thinking of you often and of ways to make you happy—please be so . . ." —LUDWIG VAN BEETHOVEN

This testament reveals not only a man in despair, contemplating suicide, but also a man of drama, a theatrical man whose "heart on the sleeve" emotions run deep. Here is an artist who knew that "all evil is mysterious and appears greatest when viewed in solitude, [but when] discussed with others it seems more endurable, because one becomes entirely familiar with that which we dread, and feels as if it had been overcome" (diary entry of 1817).

Despair is only one facet of this great diamond of a man. He is also capable of qualified optimism and resolute determination: "The humming and buzzing (in my ears) is slightly less than it used to be." And not too long after his testament: "I will seize Fate by the throat; it shall certainly not bend and crush me." He kept up a correspondence with a friend, addressing him with a touch of Mozartian humor as "Plenipotentiary of Beethoven's Kingdom" and "Most Excellent Count of Music." Not Mozart's banal humor, but hardly the dour pessimist as he is so often depicted. Referring to a sixteen-year-old student, he writes:

"I am now leading a slightly more pleasant life, for I am mixing more with my fellow creatures. . . . This change has been brought about by a dear, charming girl who loves me and whom I love. After two years I am again enjoying a few blissful moments; and for the first time I feel that marriage might bring me happiness."

This was not to be. No romantic episode, of which he had many, could interfere with his music as we get a glimpse of the prolific composer in a letter to a friend:

"I live entirely in my music, and hardly have I completed one composition when I have already begun another. At my present rate of composing, I often produce three or four works at the same time." To another friend, in a letter dated July 1, 1801, his exuberance overflows: "At the moment I feel equal to anything. I am composing all types of music."

This was no idle chatter. In the first two years of the nineteenth century, Beethoven completed his First and Second symphonies, six string quartets, his Third Piano Concerto, five piano sonatas, three violin sonatas, the String Quintet, the Septet, sets of variations, *The Creatures of Prometheus,* and a number of shorter works. Despite the sheer volume of his output, Beethoven was hardly a hermit, until the returning onslaught of his deafness in 1802 interrupted an active social life.

The dour, rather forbidding depiction of Beethoven in many of the busts and paintings that are so proliferative shows the angry Beethoven who tolerates no slight, real or imagined. His running argument with the Imperial Court was interrupted only by obsequious requests that served only to increase his seething resentment at the position in which these requests placed him. Some may postulate that his anger, never far from the surface, probably nourished some of his towering, "serious" music. After receiving rejections of his many approaches to the Court for a permanent Court position and denial of his request to use the Court theater for a concert, he wrote what was then a dangerous response:

"There are rascals in the Imperial City as there are at the Imperial Court."

Money Troubles

I wonder if Beethoven's liberal political opinions had something to do with his running eruptions against the Court and even against some of the aristocratic patrons who genuinely befriended him. Not limited to the Court, Beethoven quarreled with his devoted Viennese publisher, Artaria and Co., inaugurating, in Solomon's words, "The series of legal entanglements which drained his energies (and worked off his aggressions?) during the next two decades. Despite his later admission that he himself had corrected Artaria's proofs, he accused the firm of having stolen the Quintet; and in February 1803, Artaria filed a court petition demanding a public apology. Beethoven stubbornly refused to issue a retraction, however, even in the face of a court order."

After all, Beethoven was a child of the French and American revolutionary wars, the period of populist uprisings. *Liberté, egalité, fraternité* were sentiments that reverberated throughout Beethoven's life and in his music. I believe that it can be demonstrated specifically how Beethoven's revolutionary attitudes show in his compositions. Though some would disagree, I believe that Beethoven's unrelenting views of virtue, of integrity, and of his own personal worth were driving forces that pervade much of his music.

After one of Beethoven's main "guardians," Prince Lichnowsky, meddled in one of Beethoven's love affairs and worse, and compelled Beethoven to make changes in *Fidelio,* Beethoven refused the prince and princess entry to his lodgings. Later, Beethoven became angry at the prince's request that Beethoven perform for a group of French officers, calling it "menial labor." A Count Oppersdorff came between the two fighters when Beethoven "picked up a chair and was about to break it over the Prince's head." Fortunately, the rupture was healed, and a few years later, the prince visited Beethoven to quietly watch his protégé at work, and then he meekly departed.

The withdrawal of the annuity from Lichnowsky began a series of financial and other hardships accompanied by humiliating pleadings that occasioned these remarks: "I shall never come to an arrangement with this princely rabble connected with the theaters," and, "The thought that I shall certainly have to leave Vienna and become a wanderer haunts me persistently."

Adding insult to injury, this musical Job had still another misfortune when the currency was devalued, reducing his income by about 60 percent. But despite his persistent problems, or perhaps because of them, Beethoven continued to produce one great work after another, his Mass in C, the Choral Fantasia (a precursor to his monumental Ninth Symphony), his Fifth and Sixth symphonies, his Fifth Piano Concerto, piano trios, the Cello Sonata, *Coriolan,* and other works. Finally, after threatening in earnest to leave Vienna, waving before their eyes an offer from King Jerome Bonaparte to be his *Kappellmeister* at a salary of 600 ducats, the archduke agreed to pay him the sum of 4,000 florins annually.

Ever Amorous

Now Beethoven, feeling free to marry, writes to a friend:

"Now you can help me to look for a wife. Indeed you might find some beautiful girl . . . and one who would perhaps now and then grant a sigh to my harmonies."

Again the composer was disappointed as one after another of his series of short-term heart's desires either rejected him, just wanted to be a "friend," or wanted to use his position to better their own agendas. The only long-term relationship, between Beethoven and his "Immortal Beloved," apparently existed by mail alone. Well-known to historians by the legacy of voluminous letters, the enigmatic woman in this relationship was never identified, though speculation on the riddle of who she was persists to this day. Expert opinion has settled on a married woman, Antonie Brentano, whose husband was a Frankfurt merchant fifteen years her senior. Unhappy though she was in this marriage, she apparently remained faithful until she could no longer resist the implorings of Beethoven.

"My angel, my all, my very self, . . . can our love endure without sacrifices, without demanding everything from one another; can you alter the fact that you are not wholly mine, that I am not wholly yours? Love demands all. . . . My heart overflows with a longing to tell you so many things—Oh—there are times when I find speech is quite inadequate— Be cheerful and be forever my faithful, my only sweetheart, my all, as I am yours."

Once she offered to give herself to him, however, he held off, whether for moral or religious qualms, or from fears about his male abilities, as one writer implied. Both lovers were left in agony. Beethoven wrote: "No one else can ever possess my heart-never-never-never." And: "My life in Vienna is now a wretched life—your love makes me at once the happiest and the unhappiest of men. . . . What tearful longings for you-you-you—my life—my all—farewell—Oh, continue to love me— never misjudge the most faithful heart of your beloved L."

On finally accepting his fate to be alone, he wrote: "Thou mayest no longer be a man, not for thyself, only for others."

Legal Struggles

His overwhelming need to love as well as to be loved led this complex, very human Beethoven to limitless venality as he fought his brother's widow to gain custody of his nephew, Karl van Beethoven. Beethoven lies, fantasizes to the point of thinking his fantasies are real, utters calumnies against his sister-in-law, sues and then refuses to take the court's judgment. He behaves as one obsessed. He denies his brother's will, which hoped for "joint guardianship." He claimed that Johanna lacked "moral and intellectual qualities," and at one point called her a prostitute. Even the judge was embarrassed at the behavior of this great man.

Solomon has an interesting insight: "The negative side of Beethoven's attitude toward Johanna is quite manifest. His letters and Conversation Books are filled with vitriolic and unfounded accusations against her; he reviled her with epithets and applauded when the boy repudiated her. Such uncontrolled and passionate feelings of hostility, however, are in themselves a form of denial, an attempt to stave off powerful positive impulses . . . the more powerful the manifest emotion, the greater may be the opposite feeling which it strives to keep in check." Solomon invokes Freud: "Feelings of love that have not yet become manifest express themselves to begin with by hostility and aggressive tendencies . . . later on joined by the erotic one."

Here we see plenty of evidence that Beethoven was no emotional superman. He does the things many people do when pushed to the wall, displaying the human frailties none of us is immune to. When he finally won custody, he smothered this poor child with repeated protestations of love, and paranoia:

"I had been noticing signs of treachery (Karl meeting with his mother), for a very long time; . . . on the evening of my departure I received an anonymous letter . . . which filled me with terror. . . . Karl, whom I pounced on that very evening, immediately disclosed a little, but not all. As I often give him a good shaking, but not without valid reason, he was far too frightened to confess absolutely everything. As I frequently reprimanded him, the servants noticed it. . . . Everything here is in confusion. Still it won't be necessary to take me to the madhouse."

Johanna, now convinced that no further reconciliation was possible and alarmed at the harmful effects upon Karl van Beethoven's guardianship, began a lawsuit to recover custody. After long, tumultuous hearings, charges, and countercharges, the magistrate rejected Beethoven's clearly pathological behavior and his constant appeals and petitions to various courts. Yet Karl was awarded to Beethoven and a man named Peters as joint guardians. Needless to say, the boy, now in his late teens, was the main victim of this damaging tug-of-war, pulled from one "parent" to another throughout his growing-up years. Johanna, perhaps as a replacement for her "stolen child," became pregnant by a "noted, very well to do" person who acknowledged his responsibility. Significantly, she named her daughter Ludovica.

Thus did this drama come to an end. Beethoven had plenty of emotional fuel to feed the fires of his creativity, but his feelings of guilt and remorse were not so easy to dismiss. A friend reported in 1816 that Beethoven exploded with:

"What will people say, they will take me for a tyrant!" And the following year, he quoted Schiller: "This one thing I comprehend: life is not the greatest of blessings, but guilt is the greatest evil."

These feelings continued to gnaw at him. In 1818 he wrote in his diary: "I have done my part, O Lord! It might have been possible without offending the widow, but it was not. Only Thou, Almighty God, canst see into my heart, knowest that I have sacrificed my very best for the sake of my dear Karl: bless my work, bless the widow! Why cannot I obey all the promptings of my heart and help the widow? Thou seest my inmost heart and knowest how it pains me to be obliged to compel another to suffer by my good labors for my precious Karl!!!" Three exclamation points! Beethoven's emotions ran free and wild. He was capable of all the emotional extremes, from torment to ecstasy.

Karl, now grown up, had at one point tried to commit suicide. The turbulent relationship with his uncle, the breakups and reunions, continued, but on a level where Karl was finally able to experience his independence and express his love for his uncle. Almost too late, Karl wrote: "My dear father . . . I am living in contentment, and regret only that I am separated from you."

Beethoven, in his latter years, was still known for his eccentricities. Friends and admirers summed it up: "For all his odd ways which often bordered on being offensive, there was something so inexpressibly touching and noble in him that one could not but . . . feel drawn to him." Well known music critic J. F. Rochlitz wrote, "Beethoven's talk and his actions all formed a chain of eccentricities. . . . Yet they all radiated a truly childlike amiability, carelessness and confidence in everyone who approached him." A journalist spoke of him as "an amiable boy."

Those latter years provided the world with Beethoven's most majestic works as well as the controlled violence in his late string quartets, where he breaks the bonds of all previous music. His music leads us to a noble and glorious future.

An Ageless Figure

Beethoven leaped beyond his time for two main reasons: one, his transcendent music, which I shall address in the next chapter, and two, his humanity.

Growing up during the decline of the Age of Reason, he explored religions and other belief systems. He despised the Austrian government for its rigid censorship and its network of police spies. This was a period that opened up such revolutionary concepts as freedom and inquiry into all kinds of ideas. Beethoven was fearless in expressing his feelings, calling the government "a paralytic regime" among other less complimentary phrases. His nephew repeatedly urged caution, and after one of Beethoven's loudly expressed opinions, Karl, fearful for his uncle's welfare, shouted: "Silence! The walls have ears."

Having gone through all the trauma of his deafness and the fight against his brother's widow for his nephew's jurisdiction, Beethoven entered a new period in his later years, a time of deep reflection occasioned perhaps by his earlier studies of the Eastern philosophies or simply as a result of his maturing. These latter years also unleashed his creative powers to the fullest, leaving us some of the greatest music ever written, his last string quartets, piano sonatas, the Missa Solemnis, and, of course, his inimitable Ninth Symphony.

I have gone into so much detail about this man because his life and his music are inextricably tied together, and it touches on the basic premise of this book. As a person, he encompasses the diverse characteristics of what makes up a human being in his and our time and the potential for humans in a better future. He shows what it is to be a human being in all its glory and all its ugliness. He openly responded to the wonders and miseries that roil the waters of what we call "civilization."

Luckily, this musical and human titan recorded every nuance of human feeling in his music, so that in a better day, in another age, historians, more sentient and objective than we are, may study a climactic period in the history of man. This brief scan of Beethoven the man hardly does justice to this many-sided person. My hope is that it will encourage the reader to listen in depth to his music.

> "A scorching mysticism, a passionate intuitive belief in God-in-Nature and in the moral conscience." —ROMAIN ROLLAND

14

Beethoven's Music

Until he was in his late twenties, his music followed
older patterns. The uninitiated can sometimes
mistake late Haydn or even Mozart with Beethoven's
early music, though this early music had already begun
to defy the etiquette and conventions of the classical period of his immediate predecessors. Beethoven never
hesitated to single-handedly take on the aristocratic upper class. When Napoleon Bonaparte was looked on as
the people's general, Beethoven was similarly rebellious
and self-willed. His Second Symphony was reviewed in
the *News of the Elegant World* as "a filthy monster, a
wounded dragon writhing hideously, refusing to die,
and in the finale, even though bleeding from every pore,
still thrashes about."

A performing artist tells us as much about himself in
a performance as he does about the music performed.

The same is true about the composer and his music; one can't separate the man from the music. His unshakeable opinions about humanity, his tenderness, and his fearlessness in calling for justice and a better world are all expressed without reservation in his music. Whereas many performers and composers take refuge in technical virtuosity or artificial interpretations and superficial sentimentality, Beethoven withheld nothing. He was constitutionally unable to hide anything of himself.

This reminds me of the first time I was French horn soloist at Carnegie Hall.

Naked Horn Player

As a young horn player, I had already been solo first chair of the National Symphony, the New York Philharmonic, and the Los Angeles Philharmonic. As you know, orchestral players play from a seated position, and the horns customarily sit toward the back of the orchestra. But when I was invited to appear as soloist, I realized that I would be standing in front of the orchestra. Even at rehearsals, in an empty house, I was feeling very much exposed in this hallowed hall with the orchestra behind me. Sitting in the middle of an orchestra, surrounded by ninety or more musicians, gave me a feeling of protection or cover. But as soloist there was no place to hide. Intrinsically shy and introspective, I was very nervous in anticipation of this concert, where I would perform two difficult concerti, one by Mozart and one by Richard Strauss. What would happen if my knees shook or saliva made my lips slippery, a dangerous thing for a horn player? The horn is a treacherous instrument at best, even in its normal orchestral position. What would it be like standing up under these unusual circumstances? I was very scared. What would I do?

Here was my plan: Arriving onstage, I would take off all my clothes, remove my skull and breastbones, and reveal all of myself without reservation, my skinny body of which I was ashamed and a mind that harbored all kinds of thoughts from evil to noble. Nothing held back, my secret loves, my hates, my despairs, my hopes and aspirations. What the audience saw and heard, of course, was a young man in white tie and tails, playing his heart out. The performance was flawless, according to the reviews in the New York press, and the enthusiastic audience brought me back for many bows.

Can I dredge up this kind of total openness every day? Of course not. Any of us would get in trouble should we say everything we are thinking or act on every impulse. Making of music is a wondrous exception. Here we can and must reveal all of ourselves when we make music.

Why Perform, Anyway?

Some seasoned actors and performers have admitted to extreme nervousness and palpitations before a performance. What is it that motivates people to put themselves on the spot, facing possible humiliation, fear, and cold sweats by appearing onstage, putting their artworks on exhibition or even expressing themselves in class or in small groups? The phenomenon extends across the board.

When I was a small boy, my father took me on a trip to the small coal and steel towns in western Pennsylvania, where he sold wholesale shoes to the proprietors of shoe stores. Once, at the end of the day, too far to return home, he took me to a movie house, where upon entering we saw a raffle taking place. The MC was at the point of asking for a volunteer from the audience to turn the wheel and pick out the winner's number. Much to my surprise and mortification, my father rose from his seat, and this tired, sick old man walked to the stage, mounted the few steps, and proceeded to fill his role, his one experience under the bright lights. Did I get my itch to perform from him? Is there a "performance gene"?

When I was in my early days as a conductor and still playing the horn, I was frequently asked to play a horn concerto, usually one of the Mozarts, while conducting the orchestra simultaneously, so, of course while facing the audience in front of the orchestra, I sort of had to conduct with my body and mostly with my rear end. Once, I got confused and had a total memory lapse. I stopped the orchestra and addressed the audience, telling them that it was my mistake, not the orchestra's, and that we would start over from the beginning of that movement. The applause was deafening.

I've witnessed that same phenomenon with seasoned, famous performers in the Aspen Festival (where I was the youngest faculty member) who had the same memory lapse and got the same audience reaction. I guess audiences enjoy seeing such performers as human. Then there was the time when I was conducting Beethoven's "Emperor" Piano Concerto in Havana, Cuba. The soloist, as great an artist as any in the world, jumped a couple of bars, and I had to quickly signal the orchestra to also skip a couple of bars to be with the soloist! Except for a few cognoscenti, the audience didn't even notice it, and we came to a thunderous end of that great composition with a stand-up response.

I had the privilege of performing and recording the Grieg Piano Concerto several times with the magnificent artist Arthur Rubinstein. Backstage, he would always pace back and forth, his hands clasped behind his back, repeatedly muttering in his Russian accent: "I'm nervous, oh so nervous," and he would walk out onstage, sit down at the piano, and play like an angel!

For Beethoven, music was not just hauling out temporary openness for a few hours of performance, or a page or two of composition. For him it was a way of living every note and every phrase, no withholding, reaching into every nook and cranny of his emotions and expressing what he found in his music. We lesser mortals live out our lives in a society accustomed to polite circumspection. We close our minds and become accustomed to bodily rigidity, whether we are aware of it or not. We cover our bodies with clothes and our minds with traditional modes and superstitions. Too often we hide our truthful ideas in accepted parameters of perception. Unlike the camel, we don't stick our necks out.

Here, I believe, may be the one reason why Beethoven towers over any other composer in history. His music is not only beyond compare for its structure (and structure breaking!), but his music is prescient as well. It speaks to us today as much as it ever did, predicting a new man in a new society. It tells us the truth about life and who we are. It ranges from the sublime to devilry. And it also illustrates music's accord with the fundamentals of physics.

The Physics of Beethoven's Music

For illustrating music's relationship to physics, few composers answer the question better than Beethoven. Examples of the concatenation of physics and music can be found in other cultures and other music, but none are so clear and graphic as in Beethoven's music. For all its enormous output, variety, and complexity, his music can boil down to Pythagorean or Keplerian simplicity, as simple and elegant as $E=mc^2$. From his earliest works on, and for all its breaking with traditional forms, Beethoven's music is superbly geometrically structured.

> *"Beethoven starts with a portrait of Chaos, and he crowns the Symphony with the drama of the close of man's history, in the Elysian state of civilization."* —OTTO BAENSCH

Baensch, a German philosopher, was referring to the Ninth Symphony. On a subjective level, my scenario for his music, and particularly his Symphony no. 9, has always been a picture and a revealing of the fundamental makeup of the universe. Let us look at the basic building blocks of this composition.

Beethoven opens the curtain quietly, with the very simplest of materials, open fifths. Even before I begin conducting this work, my mind travels to a pre-universe in the ultimate quiet just before the dawn of time and space. The energy gradually builds (Beethoven keeps adding dynamics and instruments) until, suddenly, a thunderous burst of fifths and octaves crash through the emptiness. The universe is born!

Though the Symphony no. 9 is in D minor, Beethoven opens with A minor, the fifth of D minor. The only more basic interval in music (and perhaps in the universe, according to Pythagoras and the harmonic series) is the octave. Beethoven wastes no time in getting to the octave, on page 2 of the score, wherein the A is employed and quickly reaches its proper role, that is, it resolves to the basic key of D minor.

If you follow the score, or if you listen carefully to the music, you will discover the harmonic series, a relationship of tones based in nature (which we will explore in detail later), throughout this seventy-four-minute work. Beethoven's creative genius and craft require few variations, permutations, or modulations. He accomplishes the miracle of this entire work using the simplest of harmonic materials.

If you, the reader, do not read music, have no fear. I have no intention of proceeding with a musicological analysis of this or any other work. I simply want to use the materials of music to show parallels between music and physics and sociology. I'm focusing mainly on Symphony no. 9 because it "says" in the music itself what I'm trying to elucidate in this book. It addresses music's potential role in society. It is elegantly explicit in the use of Schiller's poem in the choral movement. Its musical notes follow a path as beautiful and logical as the calculus of physicists' equations.

Schiller noted: "To arrive at a solution, even in the political problem, the road of aesthetics must be pursued, because it is through beauty that we arrive at freedom." Beethoven's teenage reading of the Schiller poem "Ode to Joy" began his thirty-year obsession with the notions of brotherhood and freedom. He dreamed of liberation from rigidly repressive customs and injustices. Who has not been profoundly moved by his determined clarion call for the unity of all humankind—

"Alle Menschen Werden Bruder" (all humankind shall be as brothers and sisters)!

Beethoven was haunted by "Ode," struggling to build a musical architecture grand enough to contain Schiller's message. The germ of the Ninth is in Beethoven's sketchbooks, in his operas, *Fidelio* and *Leonore,* and in his Choral Fantasy for orchestra, chorus, and solo piano. Finally, after some thirty years, he realized his dream in the full-blown theme of the Ninth, considered by some to be the greatest ever in symphonic literature.

"The Ninth Symphony is intended as a musical narrative, a cosmic history told by evangelist/narrator, recalling the span of universal experience."

—MAYNARD SOLOMON

Beethoven's Ninth realizes the questions humanity asks of itself. It unveils unlimited tragedy and provides the ultimate, ecstatic answer in "Ode to Joy"—human unity! It is no wonder its theme has become one of the best-known and most popular melodies ever, sung in concert halls, churches, on television, and all over the world. The Ninth Symphony is the official music of the European Community (EC). It is played all over Japan every New Year's Eve, and it has been obscenely regurgitated in television commercials.

In the words of the incomparable Dag Hammarskjold, former Secretary General of the United Nations, "Beethoven has given us a confession and a credo which we, who work for the United Nations, may well make our own . . . to practice tolerance and live together in peace with one another."

Beethoven's musical passion flows into his letters, though words and structure hardly compare with the meticulous care he gave to his composing. He would compose many "rewrites" before being satisfied. It is revealing to read, in his notebooks, how he went over and over his notes, redoing passages, changing and revamping works until they arrived at the point where he and we recognize their just-rightness. His earlier sketches and drafts display all the warts and inconsistencies of his, and our, life's struggles.

Beethoven wrote some of his best, most enduring works while almost totally deaf. There is the well-known story of his conducting the debut performance of the Ninth Symphony. With his nose buried in the score and unaware that the orchestra had come to the end of the music, with the audience wildly applauding, the concertmaster had to rise from his chair, come around to Beethoven, and gently turn him to face the audience so that he could *see* the explosive applause.

Unlike music that is composed according to the passing fashion and becomes irrelevant when times change, Beethoven's music has a long "shelf life." A good part of why this is so has to do with his message and his craft. His message somehow seems to address the needs of people throughout the entire history of what has been amorphously labeled "Western civilization."

Beethoven sounds like a communist (though Karl Marx had yet to be born) when he wrote: "I am really an incompetent businessman who is bad at arithmetic. I should like things to be differently ordered in the world. There ought to be a market for art where the artist would only have to bring his works and take as much money as he needed. But, as it is, an artist has to be to a certain extent a businessman, and how can he manage to be that? Good Heavens—again I call it a tiresome business."

These tragic letters were written long before he completed the main body of his oeuvre, much of it heroic and optimistic. It has occurred to me that Beethoven's greatness and his undying popularity lie, among other things, in the fact that he reaches human depths and heights of feeling and emotion rarely reached by others. His ability to put these in his music makes all of us who listen to his music feel a little more human.

Beethoven, The Story Writer

Beethoven's giant stride crossed easily over both the Classical and Romantic ages in music. Webster's defines "*roman*" as a novel, "a type of *metrical narrative* developed in France in the Middle Ages." The Oxford English Dictionary takes it back to the ancient Roman Empire. The word and its siblings take almost fourteen columns in tiny print. Other definitions include: "When the wide field of possibility lies open to invention," and "When there is the least possible plot." Another favorite of mine is "The last romance of Science . . . is the Story of the Ascent of Man."

Here, perhaps, lies a key to Beethoven's enduring popularity. His music tells a story, and the reason he is so loved and admired universally lies, I think, in the huge span of his emotional utterances, all within the rigors of mathematical structure. Just put in your request: Do you want stormy music? Gentle, pulling-at-your-heartstrings music? Love music? Passionate music? Country music (of his time of course)? Defiant music? Death music? Dance music? Exuberant, irrepressible music?

Beethoven covers practically every facet of what it is to be human. Listeners to Beethoven's music are somehow made to feel more human. Most of us confine our lives or are confined by circumstances to a narrow spectrum of human experience, but Beethoven expands us. We can breathe more deeply.

15

Fantastique: Hector Berlioz (1803–1869)

While Beethoven set the bar at heights never scaled by other Romantic composers, there were quite a few who aspired to approach that high pitch. One of these was a young composer legatee of the Renaissance and the French Revolution of 1789. Hector Berlioz's French version of Beethoven's passion revealed itself in torment and rage. Like Beethoven, his music reflected the revolutionary spirit. Though they are alike in that they both broke away from traditional musical forms, their lives and musical signatures could not be mistaken for each other.

Berlioz was raised in a peaceful province in comfortable circumstances at the same turbulent time that Napoleon was empire-building and, for the most part, France was immersed in war. His physician father saw to it that he had a good education in classic literature. Showing an early aptitude for music, he learned to sing and play several instruments. He never learned to play the piano, and in later years he wrote that he was glad that he had been required to compose in his head, "silently and freely."

He began composing as a boy and at fifteen tried to be published, but no one accommodated him. At twenty he had an oratorio performed and at twenty-five won his second Prix de Rome. When he heard a Beethoven symphony for the first time, Beethoven promptly joined Shakespeare as one of his two heroes.

A Romantic

Like Beethoven, Berlioz was a romantic. In 1827, he attended a Shakespearean play and was smitten with the heroine, portrayed by British actress Harriet Smithson. He went to see her and presumptuously proposed marriage. She thought the young Frenchman quite mad and, of course, refused.

For months, he vainly pursued her. (Today this would be considered stalking!) To express his frustrated feelings, he composed what was the most unusual symphony anyone had ever heard. The happy result of his real-life infatuation was his towering *Symphonie Fantastique,* completed at age twenty-seven.

I have always loved this work but felt there was something missing, a dramatic scenario to complete the story. I decided that I wanted to program the symphony with the addition of *Lelio ou Le Retour à la vie* ("Lelio or the return to life"), a rarely performed companion piece that includes a narrator (supposedly Berlioz himself) and chorus. I knew that Berlioz had intended that *Lelio* follow the *Fantastique,* but I wanted to perform it *preceding* the actual symphony. I felt that performing it after was a letdown from the climactic ending of the symphony itself. Perhaps, I thought, that was one of the reasons *Lelio,* played as an appendage, was rarely performed. True, some of the text is a bit corny and there are a few other weaknesses, but I find portions of it very engaging musically and extremely dramatic, especially if it precedes the other work.

My opportunity to mount this work finally came in the late sixties, when I was slated for a Carnegie Hall concert and a subsequent recording with the London Philharmonic.

I chose certain portions of *Lelio* to perform before, rather than after, the symphony. The combination evoked a scenario that cried out for a dramatic performance or even filming. I could now "see" the entire work in one grand image.

For me, this composition, following the Classical period of Bach, Haydn, and Mozart, ushered in the next one hundred fifty years of inward-looking, self-indulgent music and art. Beethoven the giant bestrode both periods, providing a solid basis for the dawning Romantic period, which produced many great composers, such as Wagner, Mahler, Shostakovich, and Prokofiev. It was a time of inchoate yearnings (Mahler), visions of supermen and supernations (Wagner), lush fantasies (Chopin and Rachmaninoff), and the diaphanous French impressionists (Debussy). Berlioz somehow managed to both open and close the period, though his music still sounds modern. Could this be because, aside from its unique musical merit, it resonates in a persistent drug era?

In this Berlioz story of unrequited love, he sidestepped classical rules of form. Indeed, if taken out of historical context, this and his other works can sound quite contemporary. Both Beethoven and Berlioz chose to revise tradition by blending words with musical "stories." Perhaps the only difference is that Beethoven preached a global message whereas Berlioz peered inward. Both sent their messages to posterity with magnificent eloquence.

I called Columbia University professor Jacques Barzun, a great Berlioz scholar and a distinguished writer on the history of all the arts. He encouraged me, saying that the dramatic changes I envisioned were entirely appropriate for this particular composer and composition.

With this high-level affirmation, I made the planned changes in this "Lyric Monodrama" and began pulling together the necessary forces:

narrator, chorus, characters, and, of course, the large orchestra. I cut some musical portions of *Lelio* and edited out some of Berlioz's rather sophomoric text but kept his fervent tributes to Shakespeare.

As the story goes, in music and as narrated, a young man, despairing of ever winning his beloved, takes a heavy dose of opium, experiencing delirious hallucinations. The programmatic musical imagery evokes these visions splendidly. With *Lelio* now an introduction, I created special theatrical effects to enhance the mood. First, I persuaded Carnegie Hall officials and the fire department to allow a gradual darkening of the lights until the hall was totally dark. As the hall lights disappeared, music entered in the form of slow, ominous, heartbeat-like pizzicato sounds (plucking the strings with a finger rather than using the bow) from the low strings. Low woodwinds gradually merged with the strings in very soft chords until all commingled, creating a sound tableau. The low voices joined in, singing:

"Froid de la mort, nuit de la tombe, bruits eternal" ("Chill of death, night of the tomb, eternal shrieks"), from the "Chorus of the Ghosts."

One of the movements from *Lelio* is based on Shakespeare's *Tempest,* and I found a perfect transition from there, both musically and textually, into the symphony itself. The titles of the five movements are descriptive: I. Reveries, Visions and Passions; II. A ball; III. In the Country; IV. Procession to the Stake; and V. The Witch's Sabbath.

Another romantic, for the following century, was our physicist-musician Albert Einstein.

16

Einstein the Person

*"It is abundantly clear that our technology has
outstripped our humanity."* —EINSTEIN

When I was very young, I was hard put to respond
when people asked who my hero was. In my
early years I answered, "Abraham Lincoln." During my
cynical teens, nobody filled the slot until the dawning of
my "social consciousness" when, alternately, Franklin
Roosevelt, Karl Marx, and Lenin temporarily assumed
the role. My fascination with the "long march" in China
led to Mao Tse-tung as a hero, until he too became dis-
qualified in my mind.

Throughout all these central-casting years, I always
had reservations, never gave all of myself to any of the
above, never totally surrendered. Having been betrayed
early on (or so I thought), I was very sensitive to the
flaws, hypocrisy, and sheer cruelty pervading the adult
world. Heroes faded away, one after another.

Until Einstein and Beethoven

But my loyalty and veneration for these two heroes has never wavered. They have never let me down. Albert Einstein and Ludwig van Beethoven (not necessarily in that order) occupy center stage in much of this book, not as ghosts but as my very much alive teachers about the worlds of science, the arts, and life. For me, both still bestride the earth with seven-league boots. And both share many of the same values.

As a superficial measuring stick, Beethoven's "name recognition" may surpass even Einstein's among the general public. The power of music is such that its influence easily matches and probably transcends that of science. Fortunately, they are not at odds, for each reaffirms the other. Waves are the fabric of music. Without wave theory, where would science be? Science is dependent on wave theory and, by the same token, music is dependent upon and explained by science.

What kind of people were these two colossi? How did they fit in (or not) to their societies and their times?

One of My Messiahs Is Born

Albert Einstein, born into a liberal Jewish family in Germany, devoured the Old Testament until, even before his bar mitzvah at age thirteen, he happened upon books on science. These convinced him that biblical stories could not be true. Still a young child, he taught himself mathematics and music. A high school dropout with unremarkable grades in college, he had considerable difficulty in getting a job. Finally, in 1901, he obtained a job in the Swiss patent office.

Luckily for us, this job gave him seclusion as well as time to think and keep up on what was happening in the world of science. The patent office also gave him the opportunity to view the latest scientific patents and blueprints. During this period, about one hundred years ago, he burst onto the science world with a paper attempting to define the size and interaction of molecules. He thought that a universal force governed this interaction, and in 1905 he presented us with his theory of special relativity.

We have never been able to see the universe the same way again.

Einstein had a pervasive influence on millions of people long before any of us had the slightest notion about his scientific theories. His reputation as a slightly eccentric scientific genius permeated the modern world. But he also became known for his gentle humanity. Like Beethoven, he addressed issues small and large, reaching into the soul of science and man, establishing a touchstone for confirming every principle by which society and individuals judge their character.

Musician or Scientist?

Early in 1921 Einstein accepted an invitation from Zionist, biochemist, and future Israeli president Chaim Weizmann to go on a fund-raising tour of the United States and England. Weizmann was a friend of many eminent scientists. When Einstein walked down the plank from the boat in New York, he carried his violin, which rarely left his side. With his pipe and bushy hair, he looked more like a musician than a scientist. Already famous, he was greeted by the mayor, the city council president, hordes of reporters, and thousands of cheering people. His good humor and kindness only increased his popularity as he traveled from city to city lecturing at universities and attending countless meetings and fund-raising events.

Everybody knows that Einstein was a great scientist even if they know little about his theories. But fewer people know about his writings on other subjects. In fact, he was such a convincing and prolific writer that it is a wonder how he found time for his science. Through his writings and speeches on politics and issues of the day, we see that he was equally rigorous in his thinking on nonscientific subjects. Einstein had a wealth of opinion on social and moral issues.

On Human Rights

"In our times, scientists and engineers carry particular moral responsibility, because the development of military means of mass destruction is within their sphere of activity. . . . Mutual help is essential for those who face difficulties because they follow their conscience."

This is from an open letter that Einstein wrote to the Society for Social Responsibility in Science, which was published in the December 22, 1950, issue of *Science*. In 1940, Harcourt, Brace and Co. published his *Freedom, Its Meaning*: "In talking of human rights, we are referring to protection of the individual against arbitrary infringement by other individuals or the government, the right to work; freedom of discussion; adequate participation of the individual in his formation of government."

And a quotation that would probably resonate among many people in the U.S. and other countries today: "These human rights, by use of formalistic, legal maneuvers, are being violated. . . . The Nuremberg Trials of the German war criminals was based on the recognition of the principle that criminal actions cannot be excused if committed on government orders; conscience supersedes the law of the state. . . . [Our] practices have become incomprehensible to the rest of civilized mankind and have exposed our country to ridicule. How long shall we tolerate the politicians, hungry for gain, trying to gain political advantage in such a way? Sometimes it seems that people have lost their sense of humor to such a degree that the French saying 'ridicule kills' has lost its validity."

Or, "The satisfaction of physical needs is the precondition of a satisfactory existence, but in itself is not enough. Men must also have the possibility of developing their intellectual and artistic powers to whatever extent accords with their personal characteristics and abilities. Reactionary politicians have managed to instill suspicion of all intellectual efforts by dangling before their [the public's] eyes a danger from without."

Einstein recognized that expressing one's moral convictions is essential for the survival of mankind: "Those ideals, convictions, and human rights which resulted from historical experience, have been readily accepted in theory by man and have been trampled on by the same people. A large part of history is replete with a struggle for those human rights in which a final victory can never be won. But to tire in that struggle would mean the ruin of society."

On Religion

When commenting on the religious spirit of science in 1934, Einstein explained the religious feeling of the scientist: "You will hardly find one among the profounder sort of scientific minds without a religious feeling of his own. But it is different from the religiosity of the naive man. For the latter, God is a being from whose care one hopes to benefit and whose punishment one fears; a sublimation feeling, tinged with awe, similar to that of a child for its father."

Seeing religion through a broader spectrum: "The scientist is possessed by the sense of universal causation. His religious feeling takes the form of amazement at the harmony of natural law, which reveals an intelligence of such superiority that compared with it all the thinking of human beings is an utterly insignificant reflection. This feeling is the guiding principle of his life's work. It is closely akin to that which has possessed the geniuses of all ages."

In 1950, he continues this thought: "There is no room for the divinization of a nation, a class, let alone of an individual. Are we not all children of one father? The high destiny of the individual is to serve rather than rule. . . . It is nationalism and intolerance as well as the oppression of individuals by economic means, which threatened to choke the most precious traditions."

He often spoke of the difficulty of attaching a precise meaning to the term "scientific truth": "The meaning of the word 'truth' varies according to whether we are dealing with a fact of experience, a mathematical proposition, or a scientific theory. Religious truth conveys nothing true to me at all."

On Cosmic Religion

Einstein was not only deeply involved in the issues of the day but was also quite open to esoteric exploration. In his "Religion and Science" (1930) he wrote: "The individual feels the futility of human desires and aims and the sublimity and marvelous order which reveal themselves both in nature and in the world of thought." Individual existence impresses him as a sort of prison, and he wants to experience the universe as a "single, significant whole." This he called "the cosmic religious feeling."

Einstein never lost his sense of mystery, which for him became another name for God and for beauty. "The most beautiful experience we can have is the mysterious. It is the fundamental emotion which stands at the cradle of true art and true science."

In his reported last words to his son, who was at his bedside, as he picked up his equations and reviewed his unfinished statement declining the presidency of Israel, he complained, "If only I had more mathematics!" His whole life, he never stopped searching for more clues to the mysteries of the universe.

On Education

Einstein was frequently asked by education institutions to speak and write to them. On the tenth anniversary of the tertiary of higher education in America, he had much to say about education; some is germane today, when politicians are advocating for-profit education, turning education into a commercial enterprise: "All education is the culmination of the knowledge of the preceding generations. We achieve immortality by working together, learning together, and teaching each other."

In 1934, in a talk to a group of schoolchildren, he said: "Bear in mind that the wonderful things you learn in your schools are the work of many generations, produced by enthusiastic effort and infinite legwork in every part of the world. All this is put into your hands as your inheritance in order that you may receive it, honor it, add to it, and one day faithfully hand it on to your children. Thus do we mortals achieve immortality in the permanent things that we create in common."

He reflects: "Man owes his strength in the struggle for existence to the fact that he is a social animal. The value of a man should be seen in what he gives and not what he is able to receive. The teacher should be given extensive liberty in the selection of the material to be taught and the methods of teaching employed. Pleasure and shaping of his work is killed by force and external pressure."

About compensation for teachers: "In a healthy society, every useful activity is compensated in a way to permit of a decent living. It is not enough to teach man a specialty. Through it, he may become a kind of useful machine, but not a harmoniously developed personality. It is essential that the student acquire an understanding of values."

Prophetically and emphatically, he points out that "over-emphasis on the competitive system and the premature specialization on the ground of immediate usefulness kills the spirit on which all cultural life depends, specialized knowledge included. It is vital to a valuable education that critical thinking be developed in the young human being."

On His Colleagues in Physics

Theoretical physicist Leo Szilard, born in 1898 in Hungary of Jewish parents, came to Einstein for confirmation of one of Szilard's many original ideas. He and Einstein soon became close friends. From early on, Szilard's commitment to save the world was even greater than his love of physics. In 1933, seeing the emergence of the Nazis, he left Germany for London, where he worked to get jobs for the thousands of Jewish scholars and scientists who were forced out of their positions in Germany. Having read the science fiction of H. G. Wells, Szilard saw the potential dangers of atomic bombs long before experts such as Lord Ernest Rutherford declared such possibilities "moonshine." He went on to become a key figure in almost every aspect of the science and political worlds. He and his friend, chemist John Polanyi, envisioned a "republic of science," a community of independent men and women freely cooperating, "a highly simplified example of a free society."

Neils Bohr's contribution to twentieth-century physics was second only to Einstein's. After Einstein first met Bohr in 1920, he wrote to Bohr: "Not often in life has a human being caused me as much joy by his mere presence as you did." Like Einstein, Bohr was skeptical about religion, worked for women's rights, and was generally progressive. He obtained a medical degree, a Ph.D. in physiology, and was ever intrigued by profound philosophical questions. His early philosophical and psychological and social thinking had uncanny parallels with his physics, such as his watershed interpretation of atomic processes and insights into freedom of choice within the atom itself. He was fond of quoting Schiller: "Only wholeness leads to clarity. . . ." He tried to redefine the problem of free will in terms of the discontinuity of the electron. At the end of his life, he recalled his point "about the Rutherford atom, that we had something from which we could not proceed at all in any other way than by radical change. . . . That was the reason then that I took it up so seriously."

Einstein, in praise of Bohr's quantum hypothesis, averred that "the insecure and contradictory foundation of Bohr's quantum hypothesis was sufficient to enable a man of Bohr's unique instinct and perceptiveness to discover the major laws of spectral lines and of the electron shells of the atom as well as their significance for chemistry [appearing] to me like a miracle. . . . This is the highest form of musicality in the sphere of thought."

Like Einstein, Szilard, Pauli, and other scientists, Bohr took a vigorous and intense stand against fascism and Nazism. Danes were mostly anti-Hitler, and an additional factor might have been the fact that Bohr's mother, Ellen Adler, was Jewish, though nonpracticing. Bohr's student Werner Heisenberg belonged to the Hitler Youth, and during the war he contributed his scientific genius to the German war effort. Whether he stayed in Germany during the war because of his nationalism or because he was genuinely trying to steal and develop the secret of the atom bomb for Hitler's use or because he was trying to work from within to sabotage that effort has been a matter of debate ever since, as vividly portrayed in Michael Frayn's play *Copenhagen*.

Einstein never failed to acknowledge his colleagues. He had enormous respect for H. A. Lorentz, a Dutch theoretical physicist. As early as 1927 and as late as 1953, Einstein praised this great physicist and human being. He spoke of Lorentz as the "genius who moved away from Maxwell's work to the achievement of contemporary physics." He also said, "The physicists of our time are mostly not fully aware of the decisive part which Lorentz played in shaping the fundamental ideas of theoretical physics." He says that the "Lorentz transformation" was "bound to lead to the theory of special relativity," and that "whatever came from his supreme mind was as lucid and beautiful as a good work of art."

Einstein also lauded Lorentz's peace efforts for the League of Nations and other activities where he "devoted himself to the work of international reconciliation. His efforts were directed toward the reestablishment of friendly cooperation between men of learning and scientific societies. An outsider can hardly conceive what uphill work this was."

Modest Einstein gave credit to Michael Faraday, James Maxwell, and Lorentz: "The special theory of relativity was simply a systematic development of the electromagnetics of Maxwell and Lorentz"; and, "The greatest change in the axiomatic basis of physics since Newton laid the foundation of theoretical physics was brought about by Faraday's and Maxwell's electromagnetic phenomenon."

In 1930, Einstein honored Johannes Kepler on the three-hundredth anniversary of his death, calling him "a supreme and quiet man. . . . How great must be his faith in the existence of natural law to give him the strength to devote decades of hard and patient work to the empirical investigation of planetary motion. The mathematical laws of that motion were supported by no one and understood by few."

On Peace and Public Issues

"Science is and remains international. . . . The state is made for man, not man for the state. The same may be said of science."

Einstein never separated his "hard" science from his views on contemporary society. In 1954 he said, "In a long life, I have devoted all my faculties to reach a deeper insight into the structure of physical reality. To ameliorate the lot of men, to fight injustice, and to improve the traditional forms of human relations, I have expressed an opinion on public issues whenever they appear to be so bad that silence would have made me feel guilty."

He took stands on peace and war, both philosophically and in his active peace efforts in and out of the United Nations. He didn't hesitate to use his celebrated position to influence presidents, politicians, and the public. His dream of a better, more scientific society was encapsulated in a socialist society. It would be an interesting exercise to speculate on how he would view the recent world.

What would he think about a slew of articles and media interviews of television pundits and the U.S. government itself, all boldly trumpeting American imperialism as beneficial foreign policy?

The Bush administration set the tone, proclaiming its world leadership, telling the world: Do as we tell you, or we will add your country to our list of terrorist nations. Brandished over their heads was a vast arsenal of military technology and nuclear weapons, unprecedented in destructive killing capability. Just to make sure you get the idea, *The New York Times Magazine* of January 5, 2003, on a red, white, and blue front cover, blared out AMERICAN EMPIRE (GET USED TO IT) in four-inch type.

Einstein felt that World War II was an important exception to his pacifist/socialist principles, it being an important and necessary war. In a 1945 address in New York, on the occasion of the fifth anniversary of his winning the Nobel Prize in physics, he opened with these remarks: "Physicists find themselves in a position not unlike that of Alfred Nobel. Nobel invented the most powerful explosive ever known as a means of destruction par excellence. In order to atone for this and in order to relieve his conscience, he instituted his awards for the achievements of peace.

"Today, the physicists who participated in forging the most dangerous weapons of all times are harassed by an equal feeling of responsibility, not to say guilt. The wars may have been won, but the peace is not."

On the H-Bomb

Realizing the awful potential of the atom bomb, Einstein wrote a letter to the U.S. president, pleading with him to stop the research on these bombs and put an end to the incipient carnage. His plea landed on deaf ears.

In 1950 Einstein was invited to make a contribution to Eleanor Roosevelt's television program on national security and the implications of the H-bomb. He said: "The idea of achieving security through national armament is a disastrous illusion. Concentration of financial power in the hands of the military; militarization of the youth; subtle indoctrination of the public by radio, press, and schools; . . . and above all, radioactive poisoning of the atmosphere [causes] annihilation of any life on earth."

In the final years of his life, he worked strenuously to prevent the development and use of the bomb. In 1951 he was quoted in the *UNESCO Courier:* "Culture must be one of the foundations for world understanding. It is a mark of respect to be absorbed with any other's work. Respect for culture means respect for all culture, in all its forms."

Space and Time: Are They the Same Thing?

Nineteenth-century physicists thought it was absurd that physical function could be attributed to space, so they invented what they call the ether. This was supposed to act as a vehicle for light and electromagnetic phenomena. Thus, ether became a kind of matter whose only function was to act as a medium for electrical fields. By the turn of the twentieth century, the atomic structure of matter had been securely established. Ether was considered to be at the seat of forces acting across space, and since it had been understood that electrical masses in motion produce a magnetic field, it was assumed that inertia was also a field in the ether.

H. A. Lorentz thought he had discovered that the ether is fixed in space, unable to move, and that electricity is lodged in elementary particles. It was only when Einstein's special theory of relativity came along that the inseparability of time and space became understood. By the discovery of relativity, space and time were merged in a single continuum. For the first time, we see the universe as having four dimensions, now including the dimension of time.

Einstein demurs on his own special theory, which he calls "just as rigid and absolute as Newton's." He says that "theorists must be granted the right to give free rein to their fancies, for there is no other way to the goal." It is this openness that led from the special to the general theory of relativity, and later to the unified theory of relativity.

Einstein left us with some of the watermark thoughts and principles of our time—namely that space-time stretches and shrinks and is curved. He posed the theory that gravity is not a force but is the effect of the curvature of space-time, so that light and radio waves are "bent" by gravity; motion and gravity cause clocks to slow down. He predicted black holes, the final evolutionary stage of large stars that have collapsed under their enormous gravitational attraction. Einstein opened the way for the modern science of cosmology and to theories on the beginning of the universe. However, when Edwin Hubble proved that the universe is expanding, it was too much even for Einstein. He believed that the universe is unchanging, and assumed that there was some flaw in his theory of relativity. He attempted a "correction," adding a "cosmological constant" to make his theory compatible with a "static" universe. Later, Einstein called this reworking the "greatest blunder of my life," at last recognizing the fact that his theory of general relativity predicted that the universe could not be static but had to expand or contract.

Not only did he make a huge contribution to cosmology but he also had a very important influence on our ideas of the microworld, the world of atoms, protons, and electrons. His exploration of the elementary particles in the universe opened the door for other twentieth-century giants of physics to develop quantum and string theories, which we will look at in later chapters.

17

Music and Science

Einstein's passion, intensity, and humanity remind us of similar passionate expressions by Beethoven in his letters and his music. The level of passion and yearning for social change that exists today in many parts of the world is probably greater than at any other time in human history.

Both music and physics are based on consonant theories of wave frequencies. These consonants can further harmonize to affect social change. We are now familiar with the changes in physics and in music, and we can begin to see their relationship to social change. The changes in physics during our lifetimes have been as dramatic as the social revolutions. Physicists now talk of a universe without beginnings and ends, a universe (or universes!) in process and constant change. Some might add that it is about infinity. Eric J. Lerner, in his book *The Big Bang Never Happened,* reminds us: "There has always been an intimate relation between the ideas dominant in cosmology and the ideas dominant in society." The consonance of the three can lead to the "higher life" Beethoven alluded to near the beginning of this book and for which the world continues to yearn. Where Beethoven practiced the laws of science intuitively in his music, Einstein had a pretty good idea of music's relationship to science and, I suspect, to society as a whole.

How Do You Get to Carnegie Hall?

To get a better idea of how this book took shape, let us come down from the sky and land on solid ground, Carnegie Hall. More than thirty years ago, I overheard two musicians talking backstage at Carnegie Hall: "He's a fine musician, but he sure has some wacky ideas." Oops! They were talking about me.

One of the "wacky" ideas referred to was my belief in the power and place of music in society. This notion—about music's power (at that time restricted to its possible effect and use for societal change)—led me into all kinds of trouble, trouble that stretched to the highest echelons of government. But it also brought some wonderful rewards, new friends, new ideas and a richer musical experience.

A second idea about the power of music developed years later into an even more outlandish notion: *The universe is made of music.* Both ideas, which led to my writing this book, have had a never ending life, continually invigorating my personal life while gradually gaining acceptance in the scientific community as well.

Among the dozen or so physicists I approached, the majority gave cautious approval to my "theory," though I could not help but think that perhaps some were simply patronizing this presumptuous musician. However, not one physicist dismissed it. The reaction from the lay public, when hearing about the "theory," has been unanimous, something like "Sure, why not?," though some probably wondered what a musician was doing, playing around in an area seemingly so far afield from his training.

A Legitimate Theory?

My hunch, now elevated to the status of a theory, maintains that the universe is made of music. This evolved into the burden of this book— that is, the universe *is* music. I shall argue, with a little help from my physicist friends, that all is music.

One of many validations of my theory was a news item in *The New York Times* of May 16, 2001. Findings of imprints of waves "much like sound waves" that cosmologists believe were generated in the explosive birth of the universe were being made available to scientists around the world. These findings originated in Australia, using the twelve-foot-wide telescope near Coonabarabran and involving scientists at a dozen institutions in Britain, the U.S., and Australia. James Glanz, reporting in the *Times,* explained that, if confirmed, the observations would be the first direct glimpse of what amounts to a blueprint for the structure of the universe.

A few days earlier, *The Washington Post* reported separate findings from researchers in Antarctica under the bold headline "Scientists Hear Music of Universe's Birth." It tells of three independent teams of astronomers who presented the most precise measurements to date of the infant universe, as it existed approximately 14 billion years ago, exposing telltale reverberations they called "the music of creation."

My love affair with music and physics is animated by my belief that we have entered a significant juncture in human history, a time of accelerating quantitative and qualitative change—for physics, for music, and for humanity. The world-changing discoveries in physics covering the past seventy or eighty years include general relativity, quantum mechanics, string theory, and now the "many universes" theory. When three distinguished physicists, Raman Sundrum of Johns Hopkins, Gordon Kane of the University of Michigan, and Andrei Linfde of Stanford, were asked on a National Public Radio broadcast from American University in Washington, D.C., on October 31, 2002, whether these and other discoveries have an "impact on humanity," the affirmative response came promptly and without hesitation. No debate.

The effects of music on social change are equally profound, though perhaps less dramatic than those on physics. Both music and physics are based on waves and wave theory. Hence the two are closely related. The ongoing changes in these two disciplines are tightly interwoven with social change and with the effects on humanity mentioned by the physics professors. It is from that platform that I base my observations.

Over the years, after studying books on physics and talking with many physicists about relativity theories, quantum mechanics, and string theories, I came to the conclusion that in an important sense and with a much broadened definition, music is the universe. My longtime hunches were confirmed by some of the most distinguished physicists in the world. So I decided to dignify those "hunches" with the use of the word "theory." This theory was born from a recognition of the power and the universal pervasiveness of music.

This concept and definition of music has given me a broad lens through which to view the world. Seeing the world by way of a deepened definition of music together with the new physics gave me a map with which to find a road toward a better future.

Ask any group of musicians or nonmusicians to define music, and you will probably get as many answers (or confused nonanswers) as the number of people you ask. Teachers of beginning music students usually say, "Music is organized sound." Physicists would say that music consists of patterns of waveforms. Others might wonder at such a stupid question and answer simply: "Everybody knows what music is." Or, "Music is what I like to listen to," or even, "I like *good* music."

The latter response opens up another can of worms—how to define "good music." Perhaps "good music" (in my unusual sense of the term) is what this book is about. And since physics is included in this expanded definition of music, then I would say that the definition of "good physics" is that which serves the human race and the planet. Ultimately, that is also the answer to determining what is beneficial social change as against that which is destructive or retrogressive. Now we really have our hands full! Or do we? If you ask people from all around the world what kind of change they want, I would guess they would find the answer ridiculously simple: They want change that contributes to their basic needs, such as food, shelter, and a chance to educate and bring up their children decently in a peaceful world. There are some who might call trying to change the world for the better simplistic, do-goodism, or, worse, old-fashioned. Mea culpa. I take my stand.

Can physics and music provide a model for a more harmonious world? That is the tall order this book ventures to tackle. It requires a trip into the background of physics, music, and social change. It is a fascinating trip.

Facts Versus Theories

Why is a theory called a theory, not fact? Because a "theory"—relativity, quantum mechanics, string theory, etc.—has been demonstrated over and over, in experiment after experiment, and has been proven at least mathematically, maybe with a lot of help from computers. Yet the "proof," in most cases, is impossible to prove to everyone's satisfaction, especially in the quantum world, where the uncertainty principle holds sway. As in a complex music score with a simple theme, the extremely complicated relativity theories, for example, which resolved into the simple equation $E=mc^2$ (energy equals mass times the speed of light squared) has *worked*. But who is to say it would work in a different context, let's say in an expanded quantum world? Scientists are notoriously cautious, testing every tiny detail for confirmation and forever reserving a healthy measure of skepticism for new ideas. "New" in science could mean a century or two! Still, we can be grateful to the gifted writers on physics who have simplified rarified subjects, pointing out that scientific theories have supplied us with laser surgery, sonic imagery, and the atomic bomb, to mention only a few of the manifold benefits and scourges of science.

The experts would be the first to admit that the uncertainty principle, chaos, and other theories apply not only to physics but also to other fields. (Stock market gamblers take note.)

Fortunately, my main focus is on music and its concomitance and connection with all phenomena. To adequately explain my concepts of music, it is necessary to explore physics, though the reader may be grateful that I avoid the swamp of mathematics and technical language. More solid ground for me, and the heart of the book, consists of telling the story of my odyssey, the long trip that led me inexorably along a bumpy road through soul-stirring adventures.

18

The Quantum Revolution

The eighty years of quantum controversy and experimentation have yet to result in any substantial realization of the interconnections between physics, biology, psychology, economics, philosophy, the arts, morals, ethics, politics, and the macroworld we know. The uncertainty principle has held up all these years in experiment after experiment with incredibly tiny objects such as an electron, photon, or other particle. These experiments have proven that the position and momentum of an object cannot be simultaneously determined. Like the malevolent influence of a ghost, the mere attempt to measure the position affects the momentum and vice versa.

But does this principle of quantum mechanics hold true when you get many atoms in a system, such as in a mouse or a human being? Dr. Miles Blencoe, theoretical physicist at Dartmouth and author of an article in *Nature,* says, "There is nothing really in the laws of quantum mechanics that says they shouldn't apply to larger objects." Though devices such as an extremely accurate transistor device able to detect a beam of about one one-thousandth of a nanometer have been put together, problems remain, requiring improvements of several orders of magnitude before adequate testing can take place and before convincing experiments will point to the application of the uncertainty principle to large objects or, indeed, to human life. Such a physical theory would undoubtedly involve the adoption of a revolutionary attitude in real life.

Revolution in Physics

Matt Ridley, author of *Genome: The Autobiography of a Species,* writes: "Wherever you go in the world, whatever animal, plant, bug, or blob you look at, if it is alive, it will use the same dictionary and know the same code. All life is one." Darwin spent years studying the behavior of worms; he would play the piano to them to study the effects of sound and vibration. Sadly, he left no record as to their liking or critiquing his music-making.

Many biologists have written on altruism among ant and bee colonies, connecting these to Darwin's evolutionary theories, all useful and fascinating lessons for the society of human beings. More than 60 percent of human genes are basically the same as those in fruit flies, and at least 90 percent are similar to those in field mice. DNA and RNA run across all of nature—humans, animals, and vegetables. A quantum leap in consciousness remains to be accomplished, for we are truly all one, resolving, like Einstein's formula for energy and mass, into a fundamental simplicity.

A true revolution in physics, however, must include a wider philosophy able to contain various specialized and contradictory domains, then resolve them into a larger synthesis. The attempts to do so are hardly new. Einstein, whose all-embracing philosophy was able to contain a wide domain of physics, stumbled when it came to quantum mechanics. His view was that it led to an unscientific, unrealistic, and mystical approach.

Einstein's social views were quite unrelated to quantum mechanics, as were those of another German physicist, Max Planck, who in 1933 asked: "Where is science going? We are living in a very singular moment of history. It is a moment of crisis. In every branch of our spiritual and material civilization we seem to have arrived at a critical turning point." What would he add today about the state of public affairs and the general attitude toward fundamental values in personal and social life? In stunning prescience, Planck continues: "Formerly it was only religion especially in its doctrinal and moral systems that was the object of skeptical attack. Then the iconoclast began to shatter the ideals and principles that had hitherto been accepted in the province of art. Now it has invaded the temple of science. There is scarcely a scientific axiom that is not nowadays denied by somebody. . . . Almost any nonsensical theory put forward in the name of science would be almost sure to find believers and disciples somewhere or other."

Here, more than a century ago, is a feeling of collapse of the old order, together with complete helplessness and lack of understanding as to its cause. Everything is confused; culture in its broadest sense is tumbling about our ears. Now, after the World Trade Center tragedy and what looks like a long world conflict to come, the cultural and scientific worlds are reacting each in their own ways. One wonders how many wars and tragedies might have been averted—including those to come— if scientists and other leaders had combined and lent their expertise toward creating a workable world society.

19

The Science of Music:

The Music of Science

"In the beginning was the word, and the word was music. With music came dreams. Soon the dreams became visions." —ANONYMOUS

Music and physics (and poetry) are written in heightened language, and language is a social product, which indicates that music and science are also social products. Both explain us to ourselves and to the larger society. Both communicate messages of subjective and objective truth as we see it, in different periods of human history. Both delve into the awesome environment all about us, from a flower to a star.

Yes, science and the arts have much in common; both illuminate that which goes beyond the surface, both seek to see further and deeper into intricate structures, whether in the physical universe or the human soul. Both aim to simplify intricacy elegantly, whether in an equation, a painting, or a song.

Music influences human emotions as does no other art or science. From the itinerant troubadour singing a story to the sophisticated scientist, art and science have continually merged and separated, constantly recounting and reflecting the stories of our lives and times.

Though it seems, viewed superficially, that science is thoroughly objective and the arts are subjective, in fact, both reveal universal truths. Einstein was quoted as saying that science is more art than science. Physics is strongly beholden to wave theory and numbers. Music is nothing more than waves and numbers. Composers, performers, and scientists all seek "elegant" solutions to intricate problems of how the universe works, including the human part. These two disciplines are cousins, if not twin brothers and sisters.

Until Einstein, most scientists and philosophers believed that matter could be described in terms of itself, that mass, size, time, and space are objective, material qualities. Just as the audience affects the performer, so too do Einstein's theories and quantum mechanics prove that the observer interacts with and affects the material universe. This begins with the determination of the basic qualities of fundamental particles to, some contend, the stars in the sky. Amazing as it seems, even thoughts and feelings cannot be separated from the reality of matter.

Elegance in Science and Music

The supreme elegance of Einstein's $E=mc^2$ can be fully appreciated only when one tackles the web of mathematics he used in order to arrive at this simple equation. Simplicity and unity are the constant goals of scientists in their search for understanding phenomena. Music also follows these concepts of elegance and simplicity, even in what we perceive as the most complex rhythms and harmonies.

Reducing all phenomena down to the subatomic level, quantum mechanics took the next giant step—strings. Superstring theories took us down to the sub-sub-basement, reducing everything to incredibly tiny, wiggling, curved strings that, according to the theory, constitute all matter. We, in these notes, deduce that these vibrating strings—a kind of music!—are what the entire universe consists of.

In order to reach these earth-shaking developments of the past century, one must take into consideration the concept of "field." Newton tried to reduce light to motion of material points, but the phenomena of polarization, diffraction, and interference of light forced upon these theories more and more unnatural modifications until Christian Huygens's undulatory theory of light prevailed. This theory partially originated in the theory of sound, which by that time had been somewhat elaborated upon. The question had to be asked: What was the field through which light or sound traveled? This field was labeled "ether." The ether had to be assumed as a carrier of sound waves, but at that time there was no clear picture of the forces governing the ether, or of the forces acting on or between the ether, until the electric field theory of Faraday and Maxwell, which, in Einstein's words, "represents probably the most profound transformation of the foundation of physics since Newton's time." The electromagnetic field added to the gravitational field led to Einstein's *general* theory of relativity.

Einstein: The Musician/Scientist

Einstein the violinist was no stranger to acoustics. He used this science in his analysis of and solutions to problems and to the theories of physicists such as Erwin Schrödinger, in relation to Planck's constant and to energy values.

In a science broadcast in London in 1941, Einstein addressed the question of the common language of science: "The first step toward language was to link acoustically or otherwise commutable science to sense impressions. Most likely all sociable animals have arrived at this primitive kind of communication—at least to a certain degree. A higher development is reached when further steps are introduced and understood which establish relations between those other signs designating sense impressions. At this stage, it is already possible to report a somewhat complex series of impressions; we can say that language has come to existence. In childhood, individuals connected by the same language grasp these rules mainly by intuition. When man becomes conscious of the rules concerning relations between signs, the so-called grammar of language is established."

In science as in music, there is frequent use of symbols to represent abstract concepts. Language then becomes an instrument of reasoning. It is often said that if you cannot find the words to describe your thoughts, then your thoughts are not clear. I would modify this, because many of us have a clear conception or picture in our minds of the connection between things, but may not be able to find the words to describe this vision. Music is like this, too: It is a wordless language and yet gives powerful *impressions* that are reacted to in a common way by listeners.

Einstein, although he had a wonderful use of words and could explain the most abstract concepts in ordinary language, conceived many of his ideas in something like what one might call mental shapes or images or intuitive concepts. He then had the task, like all scientists, of putting these concepts into language or formulas. Thus, these concepts, in scientific language and mathematical symbols, enable the supranational character of scientific concepts. These symbols are the result of input from some of the best brains of all countries and all eras, and enable clear communication of abstract concepts.

Einstein calls this "a passionate striving for clear understanding." His seemingly simple explanation of the equation $E=mc^2$ is the following: energy is mass times the speed of light (186,000 miles per second, or 670 million miles per hour) squared. Energy can attain an enormously high number, even in ordinary objects. This does not show in most matter because that energy is only the *potential* energy, which needs to be awakened (such as in the atom bomb).

Curved Space

An expedition to observe the sun's eclipse in 1919 confirmed Einstein's theory on curved space. Colleagues observed the eclipse from opposite sides of the Atlantic Ocean and confirmed Einstein's prediction, giving Einstein instant fame. The December 1919 edition of the *Berliner Rustingerzeitung* blazoned the following: A GREAT NEW FIGURE IN WORLD HISTORY, ALBERT EINSTEIN, WHOSE INVESTIGATION SIGNIFIED A REVISION OF OUR CONCEPTS OF NATURE AND OUR EMPOWERMENT WITH THE IN-SIGHTS OF A COPERNICUS, A KEPLER AND A NEWTON. *The Times* of London issued similar proclamations. However, it took decades before the full significance of his discoveries could be applied and explained. The experiments he performed to test the many predictions of general relativity enabled us to test accurately the radio waves from quasars and from very remote objects in the sky, which in turn enabled, for example, the space mission to Mars.

Mass is now known to be just another form of energy, not unlike nuclear, heat, kinetic, and electrical energy. Rising out of $E=mc^2$, scientists can compute the exact amount of energy that can be produced by converting a given body of matter into energy.

Einstein's concepts (compositions?) are elegant indeed, comparable to if not greater than those of Bach, Mozart, or Beethoven.

20

Superstring Theory

On September 22, 1998, I read a page-and-a-half article in *The New York Times* Science section called "Ultimate Theory." It was especially intriguing to me since it reported on superstring theory and featured Dr. Edward Witten, a mathematician/physicist based at the Institute for Advanced Study in Princeton (Einstein's old home). Witten had astounded string theorists at a 1995 conference at the University of Southern California with a paper on what came to be called "superstring theory" or the "second superstring revolution." Witten theorized that the universe at its most basic consists of incredibly tiny strings in loops that require as much as ten or even eleven dimensions. Gone was our familiar three-dimensional world (four if time is included).

An intense but quiet-spoken man, called by his peers "the colossus of physicists," Witten provided a giant step in our understanding of the universe with the second super-string revolution. Though still youthful, he remains the elder statesman, the maestro, for this revolution. Ed called superstrings "a piece of twenty-first-century physics that had fallen into the twentieth century and would probably require twenty-second-century mathematics to understand."

Witten was passionate about this theory: He thought that when it was eventually understood, it would go deeper than general relativity, deeper than space and time, even deeper than quantum theory. He was sure that physics would never be the same. Steven Weinberg, physicist at the University of Texas, sees the theory from an aesthetic perspective: "Something as beautiful as string theory is rarely if ever wrong."

I was delighted to see Witten's acknowledgment by the press, though it did not come as a surprise. Years before, Nobel Prize winner Val Fitch, then head of the Princeton physics department, had described Witten to me as "smarter than Einstein." Dr. Fitch was not a man easily given to exaggeration or praise. He demanded rigorous, scientific thinking, and had little patience for science popularizations. His remarks on these were salted with contempt. Even when his own nephew, no intellectual slouch, showed his writings to his uncle, Val's reaction was crusty and skeptical. He brooked no compromise when he detected any iota of faulty or sophomoric reasoning. I valued his assessment.

Dr. Witten and I knew each other chiefly through our mutual efforts toward Middle East peace. He, like Einstein, was deeply concerned about the state of the world, particularly concerning peace between Israel and the Palestinians. One day in the eighties, driving with Ed to New York from Princeton, he responded to my question about what he was working on by excitedly telling me about string theory and its ten or more dimensions. Bewildered yet emboldened by this brilliant scientist, I tentatively spoke of my theory that the universe is made of music. Half expecting polite derision, he thought for a few seconds and calmly responded affirmatively.

Knowing Ed and having an amateur's fascination with science, I answered a sudden whim and dashed off a letter to *The New York Times*. A week or two after I sent the letter, I heard from an editor, who informed me that the *Times* was planning to publish my letter. She added that my letter had inspired the editorial staff to inaugurate, for the first time in its history, a regular "Letters to the Editor" column in the *Times*' Science section. As promised, my letter appeared, somewhat trimmed, on October 13, 1998:

> This letter is to provide a deeper definition of music. Music was always part of my being, constant companion, best friend and solace for a troubled childhood. It was also my personal magic and mystery. Even people became a kind of music. Music is much more than well-organized notes and vibrations. My original interest in relativity theory stemmed from my stretching the theory to include social relations and to all of society. In my pseudo-scientific, socio-musical lexicon, Beethoven's 9th Symphony became fact: "Alle Menschen werden Bruder" (all humankind shall be as sisters and brothers). Indeed.
>
> —JOSEPH EGER, Music Director/Conductor,
> Symphony for United Nations

Several days after publication of the letter came a call from a Penguin editor, David Groff. He said that he and his fellow editors were excited by my *Times* piece and asked whether I would be willing to write a book.

"What kind of a book?" was my surprised response.

"A book on music and physics."

"I can't write a book on music and physics; I don't know anything about physics" (a slight exaggeration but close to the truth).

He insisted: "You know about music, don't you?"

He came to my apartment and hooked me. If nothing else, this was an opportunity to share my many exciting experiences proving the power of music.

21

Race Toward the Goal

"Hearing the Echo of Earthly Music . . . Science Again Discovers How Much We Resemble the Rest of Creation"
            —THE NEW YORK TIMES, January 17, 2001

Of all the many human endeavors, science and music are the most universal. Both manage to cut across barriers of nation, state, race, religion, and economic class. And both have the potential to lead us out of the inferno of wars, nuclear detritus, radiation, and social upheavals threatening our glowing spot of green and blue in the vastness of dark, cosmic space.

Controversy has always dogged the worlds of science and music, but never so bitterly as in the twentieth century. Though scientists have a history of working together across national, racial, and religious boundaries, they are not immune to the competitive and other social toxins of our time. Einstein's general relativity theory of gravity and space-time temporarily silenced many of these arguments, certainly in cosmology—the world of the very large. But the quiet was not to last. Once quantum mechanics, the theory of the very small, showered its sparks on the scientific stage, it was like the personal and vocal electricity of soprano Maria Callas, who, upon entering from the wings, immediately and completely dominated Carnegie or any other hall she graced.

Einstein tried to disprove the basic tenets of quantum theories, but this upstart theory remained the arrogant teenager challenging the wise father. Both camps, relativity and quantum mechanics, raced toward the same goal: a comprehensive theory that could wrap up and explain all phenomena in the universe. They are still pursuing this "grand unified theory," which scientists call GUT. Einstein had such an impressive start in the race that, after all these years, Einstein continues to feed the ambitions of a wide variety of scientists and artists, including vaudevillians, novelists, psychologists, and magicians as well as physicists. Writers, looking for the next sensational revelation about a great person, rake up Einstein's torrential family life. Physicist Joao Magueijo hopes to make his mark by challenging Einstein's unshakeable theory of the speed of light with his *Faster Than the Speed of Light.* Though the scientific community is largely skeptical, the media loves the story, especially since Mr. Magueijo's image is attractive on television. Lynda Williams, who teaches physics by day, turns into the Physics Chanteuse at night, belting out original songs, incorporating the life and loves of Einstein, while exploring sex, politics, and unified theories. In her *Cosmic Cabaret* she sings about black holes, relativity, and quantum gravity. Bob Friedhoffer is a science instructor who demonstrates principles of time travel, cosmology, and quantum theory with magic tricks. He has performed at the White House, the New York Hall of Science, and the American Physical Society.

String Theory Enters the Race

"It's beautiful, wonderful, majestic—and strange if you like—but it's not weird." —EDWARD WITTEN

In the 1980s a new boy in town entered the race, more as benign referee than hostile challenger. Claiming to be the "theory of everything" and using ten or eleven spatial dimensions, string theory hopes to unify all the forces in the universe by employing advanced mathematics and abstract thinking. But due to the obstinate refusal of the theory of gravity to join the theory of the other three forces (magnetism and electricity had already combined into electro-magnetism), physicists and string theorists still had their work cut out for them.

In the meantime, scientific controversy hit the popular media. The general public gobbled up science fiction (and science!) about "big bangs" and space travel, and discussions of creation and cataclysmic theories on how and when everything would end. All this fattened up popular science writers, who in turn fed a steady stream of lurid stories to newspapers, television, and film, with plenty left over for video games. Theologians, unable to tear down scientific fact, entered the fray by adopting the new science to bolster their own brands of creationism. As religionists embraced the new ideas of science, they countered their centuries-old enemy (science) with wholesale canonization of saints and the testing of religious artifacts by way of exhaustive scientific inquiry. Thus were "miracles" and coincidences explained and judged. Philosophers and poets published theories about the theories and then morphed into television pundits.

Science's first cousin, music, was also embroiled in controversy, starting with the beginning of the twentieth-century and lasting to this day. As a musician, I was involved in these arguments, though it was not long before I came to realize that these two disciplines were allies in what was my third great interest, a science of society. Could there be such a science? Burning inside me was the need to address this science of human society, its history, current problems, and potential future. This turned out to be the most difficult science of all: what to think, what to do in this jungle of competing ideas and groups, systems and governments. I was especially troubled by the misery and injustice I saw all about me. But as the relationship of music, science, and society became clear to me, a magnificent new world opened up, one that I continue to try to explain to myself and to others. This is the daunting and exciting task I have assumed in this book. Yet, what other goal can match in importance the attempt to explore the meaning of it all?

Since the birth of the sun, evolution has been leading to this critical point in the evolution of humankind. Can we contribute consciously to the future of human evolution? Science for man is concerned with the application of knowledge for improving the human condition, or the quality of life. —JONAS SALK (paraphrased)

My purpose for these explorations is not strictly to ruminate on abstract philosophical/theological ideas, fun though that might be; it is to join in the search for answers to the paramount, crying need of our time—a better, more rational arrangement of society wherein we humans can better relate to one another. The damage that current society and its leaders are inflicting on our human species and the earth, our home, is so alarming that any occupation other than fighting to make things better fades into insignificance by comparison.

Simply as a matter of survival, it is essential for us to understand that our fate is indissolubly bound up with science and, as I shall try to prove, with music in its deeper meanings. Both are inseparable from the rest of human endeavor. They cannot be rationally discussed outside their social, political, historical, philosophical, and religious contexts.

My musical background, my love of science, and my social concerns have combined to lead me into some extraordinary adventures illuminating the power of music.

String Theory, the Final Straw

"The most musical explanation of creation ever invented. . . . vibrating strings of energy, a different "note" for each different kind of particle."
—DENNIS OVERBYE, "String Theory: Trying to Visualize Many Many Dimensions of Weirdness," *The New York Times*

Writing a book, any book, especially one dealing with physics, was the last thing on my mind. My knowledge of physics was limited to a light brush with elementary science in high school and later my fascinated nighttime reading of books on relativity and quantum theories, all strictly as an amateur. However, anyone born in the twentieth century could hardly avoid the effects of science, and in my case, a curiosity about how things work—or do not work. That kind of curiosity is the mode of operations and a necessary ingredient for good musicians and certainly for scientists. Science gradually developed as a support system for my musical life and for the employing of music as a vehicle for social change. The new physics lent validation.

Long before there were such scientific tools to lend support, there was a belief in "the music of the spheres," a theory that all the heavenly bodies make their own "music," created by their movement through the heavens. Nowadays universities have special departments to explore this "heavenly music."

Our world is flooded with natural sounds. "In the beginning was the word" opens the Gospel of John, referring to a sound, the sound of a creature, the wind, an avalanche, the sound of God. Biologists suggest that music, in some sense, may predate humans altogether.

Ancient Music: Whales and Other Singers

Whale "music" has been heard, examined, and analyzed for centuries by fishermen, musicians, and scientists. Whale and human music have much in common; whales use rhythms similar to ours and structure their songs in the familiar ABA form (statement of the theme, exposition or counter-theme, and return to a modified original theme). Despite their huge tessitura (range) of more than seven octaves, they limit their "tunes" and musical intervals to those similar to ours. Phrases, duration and use of percussive sounds, tone and timbre—these are also similar to ours. Plants too respond to music. I've watched with amazement as a plant did a dance (a very slow one!) to music. Could this phenomenon explain how some people have a "green thumb"?

On January 9, 2001, *The New York Times* Science section published

"Sonata for Humans, Birds and Humpback Whales," by Natalie Angier. She stated that "recent analyses of the songs sung by birds and humpback whales show that the animals use rhythms, notes, and harmonic patterns that are found in human music." Even more significant is her revealing that recent discoveries of musical instruments in France and Slovenia date back to 53,000 years ago, "more than twice the age of the famed Lescaux cave paintings." In fact she speculates that "the 'music instinct' long antedates the human race, and may be as widespread in nature as is a taste for bright colors, musky perfumes, and flamboyant courtship displays."

Scientists speculate that the sounds of birds, whales, and other animals are used to express territorial boundaries and sexual expressions of attraction and dominance.

In Japan, Dr. Masuro Emoto conducted an experiment in which the images of music, words, plants, minerals, prayers, etc., were captured in water in two glasses from a Japanese lake. After the water was exposed to one of the various "energies," such as music, the water was frozen and the crystals were examined and photographed. Each exposure resulted in unique, harmonious crystal patterns. In another experiment he photographed the original pattern created by the polluted water of this lake. Then he had five hundred people pray over the water for one hour, and the result was a clearly changed crystal. He believes this constitutes "an Energy Essence."

Is this quackery, or is there an important message here? The first criterion of scientific judgment is plausibility. Another is originality. Can the power of music and its easily observable effects on animals and plants as well as other objects be labeled science? "Music," it is commonly said, "soothes the savage beast" or, as some adults would complain, increases youthful savagery.

Look at any music audience—in Carnegie Hall or at a pop concert. Clearly the music has an *observable* effect. Can the effect be measured? Can it be subject to experimentation in a laboratory or on a computer? Can peripheral questions be studied and quantified? Can we quantify the ultimate effect of music on the players? On listeners? I'm often told that conductors live to a very old age—I usually respond, "I hope you are right!" Maybe someone should do an empirical study. If proven, there would be a mad race for everyone to become a conductor!

Yes, music is universal, though its forms and functions vary widely. I knew of one case where anthropologists discovered a primitive tribe that had never heard music—Western or Eastern—and had no music of their own. The two anthropologists played recordings of various Western classical composers for the mildly curious people. For most of the compositions, the people soon wandered off. Except for one composer— Mozart! This surprised me as well as other musicians to whom I told of the experiment. Most had guessed Bach! Perhaps the form of Mozart's music was more easily accessible and had greater clarity. Was Mozart a primitive?

Pairs of birds outside my bedroom window have a song that mimics almost exactly a simple arpeggio with variations. I swear that when they are a bit out of tune or rhythm, they keep trying until they get it straight! "The Music of Nature and the Nature of Music," in the January 5, 2001, issue of *Science*, suggests that other species also have the ability to create and re-create musical language, communicating within and between species. Biologists ask: "Do musical sounds in nature reveal a profound bond between all living things?"

These and other conundrums had been revolving in my mind, but the final straw that thoroughly convinced me was the advent of string theory. Wave theory, the theory that all matter consists of waves in widely varied frequencies, should have been enough to prove the true universality of music, but being of a skeptical nature, I needed a final push over the precipice of doubt. So when I learned about string theory, the theory that the entire universe, from the very smallest particle to the whole cosmos, was also made of vibrating strings, the last iota of hesitation was removed. The universe *is* made of music!

22

Superstrings:

My Neighbor's Territory

In physics and in music, as I noted previously, classifications are no longer clear-cut; composers, performers, and physicists everywhere cross over into their *neighbors' territory*. Ours is increasingly an interdisciplinary world. A very similar situation exists in physics, although specialization in science in general has increased to the point that there may be hundreds of differing classifications within classifications. In physics and cosmology, however, even though many tiny specialties exist, scientists are fascinated (even a minority of nonbelievers) by the potential of superstring theory. String theorists view their theory as the best grand unified theory yet to appear on the scene. Many think that superstrings may provide a comprehensive theory that can explain all physical phenomena, from the motion of galaxy clusters to the events inside the atom's nucleus, and may answer questions about the beginning of time, the origin of our universe, and the existence of many universes.

A superstring is described as a string coiled in on itself, smaller than any particle or sub-particle ever conceived, and many contemporary physicists think that superstring theory is the only game in town for the foreseeable future, and *the* subject for research into the makeup of the universe. Yet, nobody has ever seen such a string. It apparently takes a super mathematician, who is at the same time a super physicist, to even grasp the idea much less to explain it.

According to superstring theory, the world is made of tiny loops of wriggling strings smaller than any particle known or speculated about. Despite not having internal structures of their own, superstrings are the elements from which everything is made. One such string is about a billionth of a trillionth of a trillionth of a centimeter. Though scientists have yet to devise a conclusive laboratory test, there is considerable mathematical evidence supporting it.

The excitement of working on a "theory of everything" has been seducing many young researchers hoping to discover this holy grail of physics. A Nobel Prize would surely accompany such a discovery. Sir Roger Penrose, a physicist at Oxford University, complains that string theory has "taken over, at the expense of all other areas."

Is the Universe Made of Music?

Strings, like those on musical instruments, have played a key role in wave theories and mathematical formulations for thousands of years. We are beginning to learn that so-called inorganic matter is not so "lifeless" as we thought. If superstring theorists reveal that all matter in the universe is based on these strings, then why not postulate that the universe is one big orchestral string section?

One note played on a violin string creates, in addition to the fundamental tone, an infinite number of higher vibrations, each at a different rate, or frequency, with a mathematical relationship to the other vibrations. For instance, the tone that occurs one octave higher than the fundamental tone is vibrating exactly twice as fast. In musical language, each of these softer tones is called a "harmonic," and all the tones that occur with one fundamental tone are called the "harmonic series." In superstring theory, the superstrings could be visualized as being in the fundamental mode, while heavier particles occupy the higher modes on to infinity. Music is based on the same vibrating waves as the tiny strings (a trillionth of a trillionth the size of an atom!) described in string theories. I am sure that some mathematician, with nothing else to do, will some day calculate how many tiny violins could fit into one atom! Can it then be claimed that the only difference between superstring theory and the common understanding of music is a matter of scale?

Consider for a moment the first violinists in an orchestra. Although most orchestras prefer that they all use the same bowing (the up or down of the bow on the string and the part of the bow employed), each plays with their own style and approach, each somewhat differently from the others in the section. Yet they are all striving for one sound as a section. In effect, they are a section of soloists, but rather than creating chaos, they make one sound. Stokowski regularly instructed *free* bowing (where each violinist would choose whatever bowing felt most comfortable). I employ this method only when I want to achieve a certain effect in certain music. A more lush sound can be the result, though sometimes precision is lost.

In *Orchestral Set No. 2,* Charles Ives depicts in music an assembling of people on a train platform, "at the end of a tragic day" (the day of the sinking of the *Lusitania*), and from the chaos of people making a variety of sounds comes a triumphant, bombastic rendition of "In the Sweet By and By." In Ives's Fourth Symphony, the violinists are actually playing different music at the same time. The result may initially sound pleasantly chaotic, but there is an order that comes out of this chaos. Perhaps Ives inadvertently had a precognition of string theory!

In string theory, the "violinists" might be performing in separate times and spaces, if not dimensions, and they might not be cognizant of one another, yet they all interact with one another.

23

Up Scale and
Down Scale

Speaking of dimensions, scale is a very important concept in physics and in life. It boggles the mind when we try to imagine the difference in the size of a string described above and any other object in the universe, from an atom to an apple to a galaxy. One atom is to the width of a millimeter as the thickness of a human hair is to the distance between New York and Philadelphia. A superstring is unimaginably smaller than an atom. Now go in the other direction, the really big macroworld of galaxies and beyond, where the scale and numbers require 10 with a string of zeros much too long for this page.

The scale of our universe covers a lot of territory. We humans have been restricted to a very tiny parameter of scale. We are only comfortable within extremely narrow limits. If our body, for example, is subjected to temper-

146]   JOSEPH EGER

atures much beyond a few degrees lower or higher than the "normal" range, we begin to lose our ability to remain alive. Even with our best telescopes and microscopes, which extend the scope of the human eye far into the cosmos and deep into the atom, our limitations of scale are stunningly apparent. We can hardly conceive the distances and the numbers.

Here are some facts to give you some notion of the size of the particles in the microworld: The mass of an atom's nucleus is one-trillionth of a trillionth of a gram. Gravity affects all matter, and its range is infinite. Electromagnetism affects charged particles, and its range is infinite. Its relative strength, compared to gravity considered as one entity, is 10^{36}, ten to the power of 36, or 10,000,000,000,000,000,000,000,000,000,000,000,000! Scientists believe they can look back to 10^{43} seconds after the big bang. That amounts to 10 million trillion trillion trillionths of a second. A powerful computer computes 1.6 billion bits of information per second, and Carl Sagan calculated the probable planets in the universe at ten billion trillion! So if you are thinking of taking a space tour, you'd better take along an extra sandwich: for each light-year you travel, your chances of alighting on a planet is one in a billion trillion trillion. When you try to compare all that to the size of an atom, you will lose your mind. These dimensions are well beyond human understanding. In these emerging theories, there are some particles so small that they are beyond measurement; we can only conjecture that they exist.

The atom's "weak force," which affects neutrons, protons, and electrons, has a range of 10^{-18}, and its relative strength is 10^{29}. The strong force, which only affects neutrons and protons, has a relative strength of 2 multiplied by 10^{37}. Its range is 10^{-15}, which makes the weak force very short indeed in its range. An atom by comparison is much larger. It is about a 10-millionth of a millimeter. The atom looks like our universe, only markedly smaller. The strong force essentially holds the nucleus together. When the energy associated with the strong force is released, it produces an immense amount of energy. The energy associated with both the strong force and the weak force has made possible the atomic bomb by way of nuclear fission and fusion.

I am tempted to propose that if one could take a violin or guitar string and imagine it sufficiently elongated (miles or light-years into space) that it would be curved, eventually curving back on itself, not unlike string loops described by physicists Edward Witten, Brian Greene, and others. For these strings would themselves be subject, like all else, to Einsteinian curved space-time in the microworld of superstring theory. Taking this conjecture far enough, and applying quantum mechanics, we could well come back to the outrageous suggestion that all sound, like the flapping wings of a butterfly, effects change throughout the universe.

Eclipsing all the above is the range (scale) and "computerability" of the human mind. We compute unbelievable amounts of information in a fraction of a second merely by being alive, all with our eyes, ears, smell, and brain. Meanwhile, we are using only a fraction of the potential of the human mind. Music is basic to that human "software." The power of music indeed!

The word "scale" has multiple uses; it is a musical term that describes a series of established intervals or spaces or frequency patterns between notes, up or down, with differing musical scales prevalent in different periods in history and in different cultures; we speak of aspiring "boomers" as going "upscale" or "scaling down" their lifestyle in an economic depression. Scale is a matter of proportion. Your doctor might ask, "On a scale of one to ten, how painful is your back?" Or she may place you on a scale to determine your weight. The dictionary tells us more than we want to know about "scale": "ladder or flight of stairs, to climb or descend, a series of regular marks along a line, ratio or proportion between a representation and an object, to climb up or over, the shell or husk of fish, reptiles and some animals' tails."

Consider the scale of time. There was a flurry of media reporting in the last days of July 1999 about the Hubble constant and the age of the universe, determined by the rate of expansion. It is commonly accepted by physicists that the universe has been expanding at an ever faster rate since the big bang. If, as relativity reveals, our perspective similarly affects time, why can we not postulate a theory that 14 billion years (the approximate age of the universe) may be just an instant in some larger concept of time, which could mean that the universe is constantly contracting and expanding. We could simply be in an expansive stage of the universe and, like a flea in a dog's fur that would see a dog as a "forest" of hair, believe that that particular stage of time constitutes our entire universe. Ours is a problem of unduly limiting our context; what a shame it would be if we allowed ourselves to continue to be so restricted.

What if we stopped the world now, just when so many wonders are being revealed by science, not to mention those in music? What kind of scaled-up music have our twentieth-century physicists wrought?

Scale can ascend or descend into infinities. David Lindley contends, in his *The End of Physics—The Myth of a Unified Theory,* that string theory is, by its very nature, free of infinities. Might the same be said about music? Music might also be described as ephemeral beyond scale—in other words, free of infinity.

A Universe in Process

There is still another way that encompasses both Newtonian quantification and the new physics. This third way involves the vision of a world in process, a world of infinity, a world of the finitude and infinitude of music. But all that comes later, so I shall begin with the old physics, which we can label Newtonian, and the new physics (add your own label) of the twentieth and twenty-first centuries.

The old physics, epitomized by religious rigidity during the lifetimes of Socrates, Aristotle, Plato, Copernicus, and Galileo, came to full fruition in the discoveries of one of the greatest scientists of all time, Sir Isaac Newton. Newton's discoveries served us well for a century or two, producing many amenities for humankind while opening the window for an abundance of scientific discoveries by dozens of scientists. These discoveries, useful to this day, were dependent on measurement. Newtonian physics sees the world in three dimensions, a geometric universe. Thinking of the universe in this way gave us a magnificent legacy, lasting well into the twentieth century, enabling Einstein and other great scientists to open our astonished eyes to a maelstrom of theories about a macro-universe, vast beyond comprehension.

Its opposite, the micro-universe, is so small that only powerful computers can measure its incredible dimensions and behavior. It burst into the twentieth century with the help of triggerman Max Planck, whose insight ushered in a break with classical physics, giving birth to the quantum revolution. This magical new physics and its radically altered view of the world led to speculation about whether the two extremes were simply differences of scale or mirrors of each other. How do these extremes, so vastly different in size, relate to each other and, more important from our minuscule point of view, how do they relate to us? Could the new physics—quantum mechanics and string theories—guide us toward the much sought after unity in science and in society? Could it lead to that dreamed of brave new world that pictures a glorious future for the arts, sciences, and our human species? Or is that glimmer of hope simply a phantom water hole in the desert?

Artists and physicists once thought that they could practice their art in ivory towers, apart from the everyday realities of living. It is easy to fall into that trap. "The Elegant Universe," a television program from the book of the same name by Brian Greene, appeared on *NOVA*, a highly promoted series on PBS. Columbia University physicist Greene engagingly hosted the absorbing, Disneyfied picture of string theory, "the theory of everything," and the science that led up to it. He explains this theory in terms of vibrating "branes," or membranes, as integral to the string formations. One could presumably go through one of these branes and emerge into another universe. Physicists amuse themselves by calling it "M theory," M to signify magic or mystery, made-to-order or maladroit, magnitude or malady. I prefer another M-word: music!

The distinguished physicists on the program presented pithy, sound-bite comments and opinions, making the whole picture—from a scientist's point of view—a bit clearer to the general TV audience. It failed, in my opinion, to truly educate the public on the implications of these scientific breakthroughs. It followed too closely the "infotainment" model of television these days, a "daffy presentation—big glops of animation and surrealism accompanied by blurpy sound effects . . . kooky graphics of dancing apples, talking heads on slices of white bread and the repeated image of a cloned cellist" (*The New York Times,* Virginia Heffernan.)

Too bad; a generation of serious young scientists are pursuing this theory. It does have a truly elegant series of equations that are rather miraculously "tuned." And it is not only the young scientists; some world-famous physicists weigh in, albeit cautiously, for they find the mathematics truly beautiful and elegant. They hope that the Fermi Laboratory atom smasher in Illinois or the much larger, more powerful one under construction in Europe will lead to some answers.

All of this could provide legitimacy to my equally "daffy" intuition of some fifty years stating that the universe is made of music. But what the program ignores or treats with too light a touch, are the roots, implications, and potential of this "musical" theory for the world today.

How different was Einstein! He saw the social implications of science. In the last year of his life, July 9, 1955, he and Bertrand Russell issued the following manifesto: "In the tragic situation which confronts humanity, we feel that scientists should assemble to appraise the perils that have arisen as a result of the development of weapons of mass destruction. . . . We appeal, as human beings, to human beings: remember your humanity. . . . If you can do so, the way lies open to a new Paradise; if you cannot, there lies before you the risk of universal death."

Could Greene and the other famous scientists have given a deeper meaning to the millions of viewers? Would PBS and its sponsors have permitted the intrusion of contemporary scientific truths?

The connections between all phenomena in the universe—including us—have become apparent as never before. Yet, though we are seemingly stuck in an ominous period of history, change is inevitable. The eminent scientists, instead of leaving it an open question, might have nudged the direction of that change. What wonders might be achieved if the extraordinary brainpower of these scientists could be mobilized to explore the extraordinary power of music and physics and their use as vehicles for social change? Instead, while their work is contemporary, their behavior is Newtonian. They are addicted to pragmatism.

But they are not alone. The old paradigms continue to guide our leaders, separating the world into "them" against "us," nation versus nation, race against race, and religion against religion. Rich against poor has attained added significance as our leaders create a new, convenient dichotomy—terrorist against non-terrorist, with the word "terrorist," like its predecessor "communist," a convenient epithet to hurl at the enemy of the moment depending on the definer's (usually a state terrorist itself) flexible agenda. A constant enemy seems to be a basic requirement of capitalism, just the opposite of science where cooperation across borders and time boundaries are the norm.

24

Newton to Einstein to Bohr to Witten: Still Seeking a Home-Run Century of Revolutionary Science

Though Einstein was the first to acknowledge that he stood on the shoulders of many previous great scientists, his contributions, one after the other, set the stage for a century that produced a stunning body of scientific advances. By far the most revolutionary were his relativity theories, which have been experimentally proven over and over again. They come as close to fact as any theory can.

Much more controversial were the quantum theories that triggered century-long studies, research, experimentation, and arguments about their use or misuse. The startling implications of these theories for everyday life and for science have been bandied about almost from the start. Religious people, philosophers, and the lay public are among those involved. These two sets of theories have upset the way we look at and measure every aspect of life. The "realities" all of us experience no longer make sense in Einstein's cosmological (macro) world and make even less sense on the other end of the scale, the quantum subnuclear (micro) world. Though we inhabit a world somewhere in between these two extremes, we are still uncertain of what constitutes "reality." Can we ever be certain again? Is certainty even possible?

Conflict Among Friends

Long before quantum mechanics disturbed our equanimity and our profound sense of uncertainty, one man, more than any other, gave us our sense of certainty. He is the other giant in the history of science, familiar to all of us, whether we encountered his work in school science classes or in our daily experiences. Sir Isaac Newton, born in England in 1642, was a mathematician and natural philosopher whose three-dimensional world served us quite well until Einstein added a fourth dimension, time. Even then, things went along rather comfortably until the twentieth century began turning the Newtonian world upside down. Yet most people today are "Newtonian thinkers." A universe in which certainty reigns would be easy to understand: you trust your eyes, ears, and observations. You can take a ruler, scale, or other measuring device and measure length, width, weight, size, and speed, and presto!—you have your answer. Or do you?

The world about us that appears rather stable and solid is not so at all. Everything is in a constant state of flux. When we sink ever deeper into the microworld of atoms, particles, quanta, and strings, we enter an *Alice in Wonderland* far more magical than Alice could even imagine. There is a constant, furious dance occurring within the molecules of your body, in a steel vault or through the earth itself. Time and space are no longer separate entities but, as Einstein described it, they constitute space-time. The basic stuff (particles or waves) we are made of can mysteriously communicate with any other stuff in the universe, especially if that "stuff" is an elementary particle. Moreover, that communication does not take as long as a telephone call or the time it takes for a computer key to place a digit on your monitor. In addled addition, our simple observation of these phenomena affects the results. Just looking or measuring has some mysterious power to change the universe. Or so it seems. What goes on here is described in future paragraphs, at least as well as science has been able to explain these mad phenomena, which appear to our limited vision only by way of mathematics and experiments.

These experiments turned Newton's world inside out. But please do not get me wrong. No one could minimize the contributions of Newton, one of the greatest scientists of all time. Prematurely born in England, in 1642, the year of Galileo's death, Isaac's poor health gave him time to build models of sundials, windmills, and a water clock. Overcoming his physical weakness, he studied math and optics at Trinity College in Cambridge, and upon graduation, he fled to the country to avoid the plague that had overrun Europe. The eighteen-month seclusion gave him time to develop integral and differential calculus, the analysis of the decomposition of light into its spectrum, the binomial theorem of mathematics, and, what he is best known for, the universal law of gravitation. Upon this foundation was built much of the superstructure of science for the next centuries.

His *Mathematical Principles of Natural Philosophy* and particularly his *Principia,* published in 1687, are among the most significant scientific books ever written. In one package, *Principia* included all previous knowledge about physics while taking a giant step beyond. Without his invention of the necessary mathematics, the great systems of knowledge in science, physics, and logic would not have been able to develop as they have. His constant testing until he got it right is at the very core of the scientific method.

The laws set forth in *Principia* describe those my generation learned in high school, and I suspect they are still in the curriculum. These three laws constitute the backbone of that body of science, which has given us everything from watches to washing machines, airplanes to atomic bombs.

The first law of motion, or the law of inertia, states: A body in motion will tend to stay in motion and a body at rest will tend to stay at rest unless acted on by an outside force (such as gravity or friction).

The second law of motion, or the law of acceleration, states: The force acting on an object is equal to the object's mass multiplied by its acceleration; that is, when an object is acted upon by a force, it continues to speed up until it reaches a certain point. This law, F=ma—a force (F) is equal to its mass (m) times its acceleration (a)—is put into practice in familiar objects such as cars, spaceships, and artillery.

The third law of motion, or the law of action and reaction, deals with the interaction between bodies. For every action there is an equal and opposite reaction.

The concept of gravity—Newton adopted the word from the Latin *gravitas* (seriousness)—has been the springboard for thousands of scientific theories, discoveries, and inventions, including Einstein's theory of relativity. Gravity has also occasioned much frustration for a large num ber of scientists who, as you read this, are still trying to understand and capture even one gravity wave.

In 1916 Einstein predicted the behavior and measure of the force of attraction between any two bodies, whether galaxies, stars, or baseballs. I suppose, if asked, he might have included the attraction between two lovers.

Despite the vast improvements that exist today in equipment and in the numbers of researchers, the search is just heating up. In 1980, Louisiana State University researchers William Hamilton and Warren Johnson, running the gravity research effort at LSU, developed a detector known as ALLEGRO, for A Louisiana Low-temperature Experiment and Gravitation Radiation Observatory, an apt name for gravity wave tones that happen to fall within the audio range. This detector's musical "bar" is tuned to 907 hertz and at best can detect a quiver far smaller than the best musical ear can hear. ALLEGRO is in operation twenty-four hours a day, seven days a week. Similar ultra scale bars have been operating around the world, the Nautilus in Frascati, Italy; the Explorer at CERV in Switzerland; and the NIOBE in Australia.

But let us return to Newton. His second book, *Opticks,* showed, by way of a prism, that light is made up of all the colors of the rainbow. These and similar studies contributed to the improvement of lenses for eyeglasses, microscopes, and telescopes. He believed that light was made of particles, which went against the then current theory that light consisted of waves. This has been an ongoing argument throughout the centuries, resolved only by twentieth-century scientists who proved that light behaved as *both* particles and waves! More on this conundrum in another chapter.

Newton's groundbreaking studies took science and the world to a new stage. Some historians credit him with the Age of Enlightenment, also called the Age of Reason due to his findings on light and especially for his use of reasoning. His experiments and axioms work, and they have practical results in the objective, "real" world. Everything worked in a clear, straightforward way, like clockwork. In the everyday world of our senses, his findings still work.

Revolt in Physics

I am tempted to add "Or do they?" But I shall resist the temptation until later chapters, for this is one of the major questions of the modern world, deserving expansion into many chapters. Quantum mechanics upset the nice, neat Newtonian world, at least on the subatomic level. With this quantum theory, all that we are sure of knowing becomes uncertain. Here is still another subject (the uncertainty principle, to be explored later) that shatters most of our dearly held beliefs and certitudes. After quantum mechanics, the next revolution was string theory, led by one of its gurus, mathematician/physicist Edward Witten.

Newton's great pivotal contributions to science continue to serve, despite this battering about. Oh! For the simpler pre-twentieth-century past of physics and, might we add, music!

25

Physics and Music:

Music Waves

The science of music can be looked at from many points of departure. One area includes the history of musical instruments; from the ram's horn used in religious ceremonies for thousands of years to the modern, scientifically manufactured trumpets and French horns, from the early twang of a single string to the superb violins developed by Stradivarius and thence to the scientific analysis that enables modern string instrument makers to solve "the secrets" of the old Italian violin makers and construct a fairly good violin for an incredibly low cost (a few hundred dollars as against a few million!). Or the fairly recent (a few hundred years) development of the organ and piano, and their compromise from the "true intonation" to the "tempered scale" almost universally used today. Still another subject could include computer music and electric and electronic instruments. Fascinating as all this might be, it would occupy another book. The more germane subject and one that is basic to all music and all physics is the subject of waves.

Early man first grew aware of waves on water surfaces. Little did he know the ramifications leading to a completely different world, eventually winding up as music, science, physics, and the atomic bomb. Wave theory has led to major discoveries, including superstring theory.

The top of any wave, whether it is of the ocean or a subnuclear string is called, as one would expect, the "crest," and the bottom is the "trough." The point between crest and trough where the water or other medium is momentarily at a level is called the "node." The vertical distance from node to crest or trough is the "amplitude" of the wave. All of this, of course, also applies to a musical tone.

Some companies, in their search to find ways of penetrating the profitable market for eliminating unwanted fat from the human body, are experimenting with the use of sound waves. They hope it will make the fat cells essentially melt away. High-frequency sound waves are already used by many plastic surgeons to break down the fat cells to make it easier to remove them by liposuction. Not quite so salubrious is the use of hypersound by the U.S. Navy, capable of producing deafening, ear-damaging 150 decibels, which has caused the death of hundreds of whales and has fatally disturbed other sea life. New York police used it to scatter peaceful demonstrators at 120 decibels at the 2004 Republican convention.

A sonic technology introduced as "Cube" by Dr. R. A. Dales combines ancient Chinese therapies with a state-of-the-art computerized instrument that employs *sheng* and *ke* cycles—two of the main ways of achieving energy (chi) exchange in the twelve channels of the body. It tests for reactivity, measuring against a "norm." It then feeds back appropriate frequencies for patient therapy of bodily stresses, toxicity, and imbalances.

I tried a similar technology with a brief treatment by Andrea Ford, using a machine called QXCI that claims ability to test and treat some eight thousand body resonances. My reactions were clearly positive.

Recording artists, producers, and engineers are well aware of the wave description called "in phase," which means that the crests of two or more waves are occurring together. When the crests of these waves do not occur at the same time, they are "out of phase."

The number of crests, troughs, ascending nodes, descending nodes, or any successive points of a wave that pass a given point in one second is called the "frequency" of the wave. Frequency is usually symbolized by the Greek letter ν pronounced "nu."

As we saw previously, Pythagoras studied the sound produced by plucked strings, discovering and noting certain phenomena associated with the vibration of the string producing it. The width of the motion appeared to correspond to the loudness of the sound; as the vibration died down, the sound grew softer. When the vibration ultimately stopped, so did the sound. Shorter strings vibrated more rapidly than longer ones; more rapid vibrations produced higher or more shrill sounds.

Around 350 B.C.E., Aristotle pointed out that as the vibrating string was striking the air, that air must in turn be moved to strike a neighboring portion, which then struck the next portion, and so on. Aristotle deduced that air was necessary as a medium to transmit sound—that sound cannot be conducted through a vacuum. This is a remarkable bit of logic considering the fact that the experimental means to prove this was unavailable during this period.

In the first century B.C.E., Marcus Vitruvius Pollio suggested that the air did not merely move but also vibrated in response to the string's vibration. The same vibratory effect can produce distortion in the shape of a liquid or gas, and will radiate outward as waves from the source point.

Harmonic Waves

Music, of course, as well as sound in general, consists of waves. Gravity and electromagnetism both consist of waves. On a nuclear level, the strong and weak forces can also be broken down into waves. String and superstring theories also describe the smallest unit in the universe as consisting of infinitesimally tiny, vibrating waves of strings. In current string theory, the length of the "string" can be infinitesimally short or as long as the universe. The sounds produced would have to be correspondingly either geometrically high or low.

In order to understand the physics of music, one must first look at the harmonic motion, or the back-and-forth movement called "vibratory motion." Harmonic motion is proportionate to its displacement from the equilibrium position. The period of harmonic motion depends on the mass of the moving body and the proportion of stress and strain. Both can be determined for a particular body and the period can then be calculated. The period is the time it takes to move from the farthest point on one side to that on the other and back. It is interesting to note that the word "period" comes from "round path" or "circle."

Probably the first period observed by man was the circular motion of the sun from one sunrise to the next. Thus, this was a means of measuring time for one day. But it was only during the mid-seventeenth century that it was possible to tell time to closer than an hour or so with reasonable accuracy. Modern science probably could have never come to be until modern time-telling devices were introduced.

Galileo

Galileo struggled to measure short intervals of time accurately and was the first to discover a periodic motion that was to be put to use for time-telling. When he was a teenage medical student at the University of Pisa, he went to the cathedral one day to pray. As he noticed the chandelier swaying in large arcs and smaller ones, it seemed to him that the period of the swing was always the same regardless of the length of the arc. He checked this by timing the swinging against his pulse. Back at home he tried to emulate the chandeliers by suspending heavy weights attached to the ceiling and letting them sway to and fro. He showed that the period of swing did not depend on the suspended weights, but only on the length of the string. A string four feet long would have a period twice as long as one with a string one foot long. Further experiments have checked all the forces that are at work, such as gravity, a medium through which the pendulum swings. We can see the relationship to any musical instrument with strings.

The same principle came to be used to measure time. Robert Hooke, who formulated a universal law of gravitation, which Isaac Newton used to verify Kepler's laws, devised a fine spiral spring driven by the uncoiling of a larger main spring that is periodically tightened by mechanical winding and used in wristwatches. In recent years, this simple harmonic motion enabled scientists to measure the vibrations of the atoms within moving molecules. They have been able to create "atomic clocks," more accurate than any previous clock. So the phenomenon that drives music has also driven the development and progress of the scientific studies that have made such a difference in our lives.

Galileo made no separation between science and the arts, as is apparent in the following letter to the Grand Duchess Christina:

> Some years ago, as your serene highness well knows, I discovered in the heavens many things that had not been seen before. The novelty of these things, as well as some consequences which followed from them in contradiction to the physical notions commonly held among economic philosophers, stood up against me no small number of professors, many of them ecclesiastics, as if I had placed these things in the sky with my own hands in order to upset Nature and overturn the sciences. They seemed to forget that the increase of known truths stimulates the establishment and growth of the arts.

Galileo was then working in Holland, which was an intellectual and cultural center. The leading bookseller in Europe, Holland published works proscribed elsewhere. The openness of Dutch society and its material well-being emanated confidence and human progress.

Bach, like Galileo, was clear in his intentions. He inscribed his scores "Help Me Jesus" or "To God Alone the Glory." He invoked the traditional metaphoric link between music, God, and the soul. Moreover, these sentiments are part of the substance of his music as well as his words. In his Mass in B Minor, the sighing violins descending into the low-pitched, earthly realm of the bass line illustrate the Incarnation of the Holy Spirit. A Bach fugue is a replication of society itself, the separate, independent lines coming together in the fugue, which sets up a community of like minds quite different from the polyphony that preceded it.

Bach's D-sharp-minor Fugue from *The Well-Tempered Clavier* is based on the relationship between the first tone of the scale (the tonic) and the fifth (the dominant). The same might be said about Beethoven's Third Symphony, but this relationship is even more striking in his Ninth. The interval of the fifth is the most consonant of intervals smaller than an octave. And it has the most overtones. Much of music gains its effect by upward or downward movement, delayed resolution, dissonance finally turning into consonance, and the breaking of symmetry and then returning to symmetry. "Leise, Leise," sung in Carl Maria von Weber's *Der Freischutz,* achieves its affecting lament simply by a chromatic, tone-by-tone downward movement. The same technique is equally effective when used by Purcell in his *Dido and Aeneas.*

Haydn's extraordinarily large output (104 symphonies, and many keyboard compositions, string quartets, and other works) is a story of development and change. In his earliest works he is preoccupied with formal problems. At first he is not greatly concerned about the significance of his structures as human experience, but his later works are precursors of Beethoven in that regard (Requiem, Symphony no. 104, etc.).

The ever moving patterns and proportions in music resemble the constructions that physicists or mathematicians create, which are not unlike the more mathematically complex music of Stravinsky. The "math" is fairly simple in his earlier works, but becomes quite complex later on, for example in *The Rite of Spring* and the 12-tone works. Even more complex are the rhythms, harmonic changes, and variation atop variation in Indian ragas. The simplistic rhythmic, harmonic, and thematic repetitions of most popular music are boring to most classical and jazz lovers.

New York Times music critic Edward Rothstein points out that "the musical theme and variations are not just one musical form: it is THE musical form. It is, rather, that which allows music to be formed. Whenever we are asked, in music, to feel some connection between one moment and the next, we are asked to attend to elements that remain the same and those that change. Variation is music's mode of argument. It establishes similarities and analogies. We know two moments are analogous when there is enough similarity between them to create a sense of recognition. These moments of analogy connect not just such instants but also the region surrounding them. Music is structurally ordered by analogy," citing the Mozart variations based on *Twinkle, Twinkle Little Star.* He then asks the ultimate question: "What gives music its power to inspire fear and dread and ecstasy in its listeners?"

Rothstein says that "Bach's fugues, though written in a religious period and by a religious man, contained premonitions of the secular enlightenment; their themes almost become objects of rational knowledge, their every gesture and interval subject to study and illumination." The art of Einstein's general theory of relativity is partly due to the ways it is harmonious with laws and concepts seemingly but not truly unrelated to it.

Max Weber noted that "the development of western music followed the course of the rationalization of . . . the displacement of religion by civil authority, the increasingly intricate structural relations between social organizations, the systematization of knowledge itself." Music, he argued, "passed from religion to science in secular life, from manipulation and repeated patterns to the exploration of hierarchy and structure, from regulatory boundaries on harmonies and intervals to attempts to create a form of abstract knowledge about combinations of sounds. Music might be considered in this light as a counterpart of western science."

26

Proof?

Though string theory has yet to prove itself experimentally, it has engaged some of the brightest minds in the field, among whom are Dr. David Gross, an early adherent and director of the Kavli Institute for Theoretical Physics at the University of California in Santa Barbara, and the aforementioned Dr. Edward Witten. Gross recently shared the 2004 Nobel Prize for Physics with Frank Wilczek and David Politzer.

Dr. Witten states, "If it turns out that string theory, which has led to so many miraculous-looking discoveries over so many decades, has nothing to do with nature, to me this would be a remarkable cosmic conspiracy."

Science writer James Glanz reported in *The New York Times* on March 31, 2001, that "Dr. Stephen Shenker, Stanford physicist, believes that the theory's strong influence on several branches of pure mathematics has provided a good test of the intellectual horsepower of the idea, and that the wider impact suggested both that physicists had not missed some crucial mathematical error and that the theory was rich enough to be an encompassing theory of nature."

Dr. Stephen Hawking is a bit more cautious: "For most of the last hundred years, we have thought that the theory of everything was just around the corner, we keep making new discoveries, but I do not think we can yet say that the end is in sight." He suggests that physicists should keep in mind that they have been wrong before and that string theory still contains many mysteries.

What with federal grants, tenured faculty positions and prestigious awards, Dr. Gross reports, "Nowadays, if you're a hotshot young string theorist you've got it made."

God's Guitar

Science writer Dennis Overbye describes the theory's elementary particles as "notes played on God's ten-dimensional guitar. All our ears could hear in this cold lonely universe in the era of galaxies and protons was the lowest bottom octave. The next higher octave and every octave after that would raise the mass-energy of these notes by nineteen orders of magnitude.

"There were an infinite number of octaves theoretically, but to hear even the next highest, physicists would have to pluck a superstring with a Planck mass of energy, 10 billion *billion* gigs of electron volts. . . . The full power of superstrings was reserved for a crack of a moment at the beginning of time, when God's accelerator was fully charged, a golden second when the symmetry of nature must have been more dazzling than a diamond with a thousand faces."

According to superstring theory, the entire universe may be said to consist of waves or strings, and some claim that the universe is in the key of B-flat. The sound of music writ large!

Dissonant Music

Both beginners and big-name physicists got into the game. Skepticism is in the nature of science. Theoretical physicist Amos Yahil, exploring the concept of cold dark matter, another major puzzle, called it "just a cute idea; don't take anything particle physicists tell you as gospel truth. They are the most theologically promiscuous people in the universe. As soon as something breaks down they will invent something new. . . . I know how easy it is."

Among the fathers of grand unification was one skeptic. Harvard theorist Sheldon Glashow argued that leaving the realm of experiment and being guided solely by elegance, the superstring theorists had left science. In *The Charm of Physics,* Glashow wrote: "Mathematics can masquerade as physics." Despite his skepticism, he was no less involved in the excitement of physics: "Once upon a time, my buddies and I figured out that there had to exist a fourth type of quark. Ten years later it was found. That feels good. It is the kind of feeling I have had a few times in my life. It is a challenge—a game. It is like going to a magic show and figuring out how the tricks are done. It is like reading a detective story and trying to anticipate the ending. It is all of these things. It is a very human activity except it is focused on one very simple question: How does it all work?"

Music and Doppler

The Doppler effect [named after Christian J. Doppler (1803–53), Austrian mathematician and physicist] has been used by scientists to measure the planets and other objects in our solar system. It occurs in everyday life as the sound of a train or ambulance as it approaches and fades into the distance, the pitch increasing as it approaches you and lowering as it fades. If the vehicle is stationary, then the sound waves remain the same and the pitch of the sound also remains constant. For great distances, more accurate calculations can be made using light waves and "color shifts" to measure the age and distance of stars. Scientists factor in color, absolute brightness, and apparent brightness.

Astronomer Edwin Hubble was thus able to measure some of the most distant stars in our galaxy. Some are more than two million light-years away. His discoveries gave cosmologists a new perspective on the scale of distance in the universe. The universe was no longer seen as finite but as infinite. A second discovery by Hubble led him to believe that the universe was static. This view was held even by Einstein at the time.

Scientists have been able to use frequencies of light to ascertain the pace that stars are moving away from us by measuring the redshift, or approaching us, by measuring the blue shift. (These are the colors displayed when an object is moving toward or away from us.) Hubble found that almost every galaxy was redshifted, giving him the conclusion that the galaxies are moving away from each other, thus the universe is expanding and is not static after all. When Einstein met with Hubble at the Mount Wilson Observatory in California, he announced that the cosmological constant had been his greatest blunder. Einstein was very moved after a scientific discussion with Hubble and with Georges Lemaître in 1931. With tears in his eyes, Einstein was reported to have said, "It was the most beautiful and satisfying interpretation of astronomical science."

Many nutty ideas have plagued scientific inquiry since Einstein's general relativity theory and especially arising out of the popular fascination over the magic and mysticism of quantum mechanics. I do not believe for a moment that suggesting the world is made of music is a foolish notion; this past century has produced such mind-boggling discoveries that they ignite the imagination of the most rigid of mathematicians. There is now ample proof that the world is not as we see it. Our daily experiences provide no clue as to how the world really works. We are forced to stretch our imagination.

Our view of the world has truly been turned upside down. We have seen a revolution in our understanding of how the nature of the universe functions in the microworld, of atoms and particles, and in the macroworld of stars and galaxies. This revolution, far greater than any in history, is still going on. The problem is that we have not been able to integrate the full meaning of this revolution and are acting as if we are still living in the world of Newton and Descartes. As important as were the contributions of past geniuses, going back to Galileo, Pythagoras, and the sages of five thousand years ago, we are finding that we cannot go home again. We must adjust our thinking and our lives to the more realistic world that the new syndrome presents us. Otherwise we will continue to be living in a dissonant, obsolete world, sliding to disaster.

Could music help? Surprising as it sounds, it is not a new thought. The one path not yet taken is via music. Music may be productive, not only in pointing the way to grand unification, but at the very least, music in all its dimensions may provide some useful directions for further research.

One of the great mathematicians, philosophers, and scientists of his and any time took that path.

27

Kepler's World Harmony
to the Rescue

*"The heavenly motions are nothing but a continuous
song for several voices (perceived by the intellect,
not by the ear), a music which sets landmarks in the
immeasurable flow of time."*
—JOHANNES KEPLER, *Harmonices Mundi* (1619)

This innovative thinker set landmarks in the history
of science. Johannes Kepler (1571–1630) was and
still is well known for introducing the laws—named after him—of planetary motion. He was possessed by the
one idea that dominated his work, the idea of world harmony. Though this idea had existed for thousands of
years, Kepler was convinced of its truth and proceeded
to demonstrate its existence in the form of musical laws.
His series of books addressing world harmony, completed in 1619, brings together and advances the premises with which others in the Baroque period were
familiar—that is, the same musical laws that go back to
Archimedes (third century B.C.E.), who figured out
geometry and discovered the means of measuring the
circumference of circles, and he came up with a number
that is awfully close to our 3.1416 . . . (π "pi"). His pre-

cise mathematical approach to this and other problems was a valuable precursor to calculus and the concept of infinity, a concept that involves struggle even today in the minds of scientists much less us laypeople. Unfortunately, his writings were lost for many centuries. And before Archimedes, another Greek, Pythagoras, gave us a body of math, with the help of music and as applied to music, which has served us ever since.

In the fifth and sixth centuries B.C.E., Pythagoreans were engaged in the mathematical beauty of geometric equations, applying these to the heavens. They believed that the planets gave forth harmonic tones; each heavenly body, starting with the moon, intoned a note higher than the previous planet. Kepler was devoted to this concept of celestial music, audible only to God.

Musicians have long seen Kepler as sympathetic to their cause. Paul Hindemith, a well-known twentieth-century composer, wrote an opera about Kepler, *The Harmony of the World.* Hindemith's system of composing conforms, though in a highly individual way (he is fond of fourths), with the relative consonance and dissonance based on the harmonic series. This series explains why atonal compositions are generally so unattractive to the public as well as to musicians: They are divorced, as Kepler claimed, "from the harmonic laws of creation."

When only twenty-three, Kepler published a work devoted to the principle of world harmony, the *Mysterium Cosmographicum.* In later works, he admitted that his planetary laws resulted from his preoccupation with world harmony, though he approached the solar system questions with scientific precision through years of calculations. He describes his approach, beginning where the Pythagoreans left off, in a letter to a colleague:

> In my harmonic researches, my best schoolmistress has been experimentation. Over a resonant chamber, we stretch a metal string. With a movable bridge placed beneath it, we travel back and forth on the string—to the right and to the left—striking repeatedly the two segments of the string divided by the bridge, and then we remove the bridge and let the whole string resound. The rest we en-

trust to the judgment of our ear. When the latter assures us that both segments make a concord with the whole string, one draws a line at that point on the soundboard and measures the length of both sections of the string. Thus one discovers what proportions emerge. Sometimes the two sections are consonant with each other, but not the whole string; at other times, one section will be consonant with the whole string, while the other creates a dissonance, both with it and with a whole.

Thus he worked out for himself a fact known since antiquity, that the intervals of music are linked inseparably to numerical ratios, usually primary ratios of string lengths, wavelengths, and frequencies; the ratio of 1-to-2 produces the octave, 2-to-3 the fifth, 3-to-4 the fourth, and so on.

Remarkably, this theory of ratios, which also plays a role in geometry, became the guide for Kepler's subsequent investigations that enabled him to finally succeed in proving his harmonic laws in the cosmos—that is, the musical harmonies generated by the planets—which he had sought all his life.

In the mix of his books, he covers astrological problems and the laws of geometry, and he was the first to use the terms "major" and "minor" in music. Without benefit of a computer, he calculated the angles of the planets, measured from the sun at their extreme points within a period of twenty-four hours. His system produces simple harmonic intervals that correspond to the musical consonances. He said, "The Eccentricities of the individual Planets have their Origin in the Concern for Harmonies between the Planets." He ascribed his discovery to

"the Creator, source of all Wisdom, constant preserver of order, the eternal and absolute fountain of all geometry and harmonics, He, I say, this divine Artisan, His mighty self has joined together the harmonic ratios . . . which expresses the idea of the spheres that surround the six celestial bodies [that were known at the time] for the purpose of regulating the movements of those bodies. From these two component parts, a unified, balanced system has to be made."

It took scientists until the twentieth century to apply Kepler's methods to the planets Uranus, Neptune, and Pluto, all discovered since Kepler's time. They confirmed that Kepler's planetary harmonies are true, even to the sounds that we can hear. A fascinating subject for scientific research, in light of quantum mechanics, would be the contribution that the human and animal ear makes to the laws of consonance, dissonance, pitch, and timbre. Such studies might be extended to more esoteric realms, as far as the cosmos and, in reverse, to the atom. One hypothesis, published by Heinrich Husmann in 1953, is that there are sounds that arise in the ear besides those that reach the eardrum. These inner sounds are "subjective overtones" and "combination tones." Our ears are truly magical mechanisms, meriting more than the limited anatomical interpretations usually given them.

The laws of harmonic relationships and ratios are seen also in biology, chemistry, chemical compounds, cell division, anthropology, geometry, acoustics, crystallography, and the list could go on.

Though the "music of the spheres" concept dates back to Pythagoras and probably before, Kepler created the formulas for his six laws of motion with scales that symbolized the six known planets of his time. Basically, each planet creates vibrations—sound waves—caused by the path of its traveling through space. Therefore, each planet produces music that stems directly from the speed of its motion in orbit. The belief persisted until Kepler's time that the planets moved in circular paths and at constant rates of speed, thereby producing one note each. Kepler found that the planets moved in elliptical paths at different speeds. Thus he changed our understanding of planetary motion, what has been called the "celestial chord." His concept of the music of the spheres held that it was a continuous, ever changing song. We may think of Earth as producing a dissonant minor second. One writer described Mercury as a piercing piccolo and Mars as singing a fast, allegro song. The combined planetary symphony may be waiting for us humans to reclaim it for its heavenly sounds, taking us to where we can hear a grand chorus and a symphony orchestra made up of stars, atoms, and planets with the inhabitants thereof!

Kepler's dreams and reputation, long demeaned, has been restored in contemporary times. At least twenty years ago, Yale University intro-duced a formal course of study in "the music of the spheres." A former colleague of mine, Willie Ruff, a fine French horn player who specialized in jazz on the French horn (not an easy combination) taught this course. His research led to the claim that certain tones and simple ratios are con-gruent with the musical laws underlying the solar system.

28

Appearance or Essence?

Throughout recorded history the heavenly bodies have fascinated man. Without the benefit of electric lights, man watched the stars and made up stories. Guesswork, prophesies, conjectures, and hypotheses followed soon after. The sheepherder was probably the first scientist, wondering about the essence of things rather than their appearance. Primitive man learned early that curiosity could save his life. A beautiful forest might harbor a dangerous animal. A prickly cactus that looks distasteful on the outside can hide a sweet inner fruit.

More than two thousand years ago, Lucretius, premier musician/poet of Roman science, mused: "To bear all toil and wake the clear nights through, Seeking with what of words and what of song, I may at last most gloriously uncloud. For thee the light beyond, wherewith to view. The core of being at the centre hid."

Shakespeare's Polonius, in *Hamlet,* declares: "I will find where truth is hid, though it were hid indeed within the centre." Karl Marx claimed that all science is founded on the inner-outer duality of a thing, the coexistence of appearance and essence. Swiss psychologist Jean Piaget wondered: "What is the essence if not logical-mathematical structure leading to an understanding of a material object?" Thus did artists, poets, scientists, and philosophers agree that one must look more deeply into things.

In the second century C.E., Ptolemy, a Greek mathematician, astronomer, and geographer, formulated a system of heavenly motion with a stationary earth at the center of the universe. His system "explained" the past and future motion of the planets. In order to resolve the discrepancies, he made small adjustments of the then known seven planets. Extremely contrived and understood by only a few, it was accepted Christian dogma for many centuries. Until . . .

Copernicus Moves the Earth

Jumping to the fifteenth century, Nicolaus Copernicus (1473–1543), a scholar and a devout canon of the Catholic Church, was disturbed by the arbitrary complexity of the Ptolemaic theory. He considered a more reasonable arrangement of the circular motion of the heavenly bodies; he had the remarkable realization that in order to have a simpler system, one had to abandon the fixed earth, and had to put the sun at the center of the universe. For putting forth such a radical notion, he quoted Cicero, who, among others, believed that the earth moved, an absurd opinion at the time. According to the Copernican system, the earth and all the planets moved in concentric circles around the sun.

The Catholic Church disagreed, quoting scripture:

> Then spake Joshua to the Lord before the children of Israel: Sun stand thou still upon Gideon and thou moon the valley of Ajalon. And the sun stood still and the moon stayed until the people had avenged themselves.

The Church declared Copernicus a heretic, and his books were forbidden as false and altogether opposed to the Holy Scriptures. This ban remained in the Church until 1835.

Where Truth Is Hidden

Man's curiosity for exploration of the many terraces and complexities of reality reached an apex in the nineteenth and twentieth centuries. In the twenty-first century we find its opposite, in a reversion to obscurantism. These opposing uses of reality collide, as science and the arts are intrinsically truthful, while politicians and fundamentalists gain by obfuscation.

Appearance and essence coexist comfortably in music and physics. The spatial and temporal elements in music are comparable with those of space and time in physics. Music consists of waves, which have elementary characteristics such as amplitude and frequency of oscillations. While this is also true in physics, wave theory is so taken for granted that its daily use, in mathematics, for example, is like our autonomic nervous system, whereby we are scarcely aware of our breathing. The two disciplines share many of the same terms and concepts, such as scale, harmony, elegance, context, et cetera. Both disciplines profoundly affect human society and individuals within society by exposing the contradictions between, for example, the appearances in astronomy and the essences in quantum mechanics. These contradictions are easily detected in music by the lyrics in song or opera and more hidden in the tensions and resolutions within the musical sound.

Contradictions abound in society, as they do in all of life. Both musicians and physicists work on cooperative effort. However, this harmonious aspect is in constant struggle with the everyday real world of competition and politics. The great Italian conductor Toscanini was vehemently outspoken about his abhorrence of fascism. Most musicians shared his views, as they do today. Socially conscious physicists include great names such as Bohm, Bohr, Witten, and Einstein. Rock stars make no bones about their opinions, expressed in their lyrics and music.

Broadway music often deals with social themes. Alfred Appel, Jr., compared Louis Armstrong to Heisenberg in *Jazz Modernism.* Next on the agenda, at least from a long-term point of view, is a connection between music and a scientific approach to society.

The disparity between our experiences in "real" life and the knowledge that we now possess about the workings of music and the universe compels one to search for alternative ways of thinking and being that might generate a musical model of being. By the same token, the new theories of physics envision a unified world. A recognition of the macro-universe and the micro-universe leads to an alternative consciousness that sees harmony, community, and creativity as the nature of nature, the nature of us human beings, and the very nature of the universe.

Indeed, this is *the* real world, and what we have come to know as the "real" world is an aberration, a perversion of reality that can be overcome and transcended. Alternative ways of thinking and being that follow the musical model are congruent with the new physics.

The new physics, I believe, supports my intuition that music, via wave and other theories, turns out to be a key to the real world. What we have come to know as the real world is a narrow, shortsighted view not only in physical, scientific terms but also in human relations and in society as a whole.

29

Modern Physics

Gravity was the subject of Einstein's 1915 paper on the *general* theory of relativity, which united inertial and gravitational qualities in an objective form. For example, in an orbiting spaceship, the downward acceleration due to gravity is indistinguishable from the outward inertial acceleration—the "equivalence principle." This manifests itself as weightlessness.

The earlier *special* theory of relativity deals with time and with the relations between different observers traveling at different relative speeds. It introduces the notion that time can be treated as a fourth dimension, in addition to the three spatial dimensions. General relativity is primarily a theory of gravitation. For this, Einstein found it necessary to introduce the idea that the four-dimensional space-time of special relativity has curvature—a non-Euclidean space in which, for example, the angles of a triangle no longer add up to two right angles.

For both general relativity and special relativity, objectivity is strictly maintained. To the end of his life, Einstein defended this position against the determined opposition of the adherers of the Copenhagen interpretation of quantum mechanics.

It was not until the 1960s that general relativity began to have extensive observational confirmation. The general theory opened up an unexpected new avenue of investigation when observations from the new 100-inch telescope at Mount Wilson in California revealed the redshift in light from distant galaxies, a discovery that appeared to find an explanation in the concept of an expanding universe, which had been predicted by general relativity. This led to an expansion of quantum mechanics to usher in modern astrophysics and the big-bang theory of cosmology.

Since the 1980s, continuing failures in attempts to extend the standard model of quantum mechanics so as to include the gravitational force have, more and more, given way to a new approach. In superstring theory the four forces of quantum mechanics and general relativity are brought together in a single theory in which the four dimensions of relativity are increased to ten dimensions with the six new dimensions supposedly "compacted" into a very small space. Mathematical experts speak of the possibility of still another, eleventh, dimension.

Time, Space, Light:
Make It Simple

I guess God was talking about space when he spoke of darkness. But then what is light? And what is in the darkness? Physicists are still arguing about what the universe, especially dark matter, is made of and how to capture, define, or measure it. We hangover Newtonians love to measure things, anything we can get our hands on. But how do we measure the speed of light? We know that in order to measure anything we have to measure it in relation to something else. The something else might be a foot-rule, your children's height against yours, or the roadway over which your car is traveling. For light, we'd have to measure it against the space through which it "travels."

In the latter half of the nineteenth century, physicists thought that the ether concept of space might define a frame of reference that would enable them to measure the speed of light as well as the motion of other objects. The shock was explosive when experiments, notably by U.S. scientists Edward Morley and Albert Michelson, showed that the speed of light is the same in all directions. Einstein pursued the matter and came up a few years later with explanations of what was then called "ether drift."

Einstein never stopped trying to make his theories mathematically and conceptually simple. He was aware, of course, of Galileo's calculations that had been accepted and tested. These calculations agreed with all experience for hundreds of years. But there was one small problem: they didn't apply to light.

He was also aware of the experiments by the two American physicists, who showed that the speed of light under different conditions of relative motion was still always the same (186,000 miles per second). But even these Nobel Prize winners never stopped being troubled by their experiments and could not understand why there was no relative speed of light.

Playing Catch-up

Of one thing, Einstein was convinced: You could never "catch up" with a light beam. At the tender age of sixteen, he conducted his famous "thought experiment." The rest is history. When Einstein was sixteen, he asked himself the question that would eventually lead to a revolution in how we think of science and of the universe: "The question stayed with me for the next ten years. The simple questions are always the hardest. But if I have one gift, it is that I am as stubborn as a mule."

He later asked: "If light were a wave, then no matter how fast it traveled, it should be possible to catch up to its peaks and valleys. But then what would I see? Would light stand still? Would time stand still? Would I ride this wave forever?"

Though the physics of sound and water waves was well-known in his time, what a leap of the imagination to apply the concept to light!

Suppose I could attain the speed of light, then ride alongside that beam of light. According to Maxwell's theory of light, it is a vibrating electromagnetic wave. Therefore, if I move along at the speed of light, I would see a wave that is stationary in space. This turns out to be an impossibility according to Maxwell and contrary to experience as well. The vibrations and oscillations of a wave are the result of a variation in space and in motion. Like watching a film, each frame in the film is very slightly different and yet you perceive that the film is in motion. If you should be able to move along the film at the same speed or stop the film, your perception of motion disappears. You can see the film only when the film is moving in respect to you. If you stop the film, it disappears. The same is true with light. You cannot catch up to light.

As he put it in the theory of relativity, a "reference frame" is the framework from which one observes. And the reference frame may be in motion or at rest. Einstein argued that you cannot go any part of the way toward the speed of light, or otherwise you leap from one reference frame to another in which light would progressively go slower until you caught up with it. This violated the laws of physics. Thus Einstein postulated that the speed of light must be the same in all reference frames. It is called the principle of relativity.

Maxwell's theory had stated that "stationary light is impossible and has never been observed." Though hundreds of years of experiment, observation, and applied physics had verified Galileo's rules and the vast framework of Newtonian mechanics—from the behavior of molecules in gas to the motion of the planets—these were rendered obsolete by rela-

tivity. Thus were Galileo and many of the physical laws, including Newtonian mechanics, challenged.

Since the 1880s there have been numerous verifications of the constancy of the speed of light, justifying Einstein's intuition and logic. Einstein continued to explore the consequences of the principle of relativity
and the invariance of the speed of light, believing that these two ideas
were fundamental postulates of his new special theory. He announced
this to the astonished world in 1905 in the classic paper "On the Electrodynamics of Moving Objects."

Uniform motion is not common in life because most objects either
speed up, slow down, or start and stop. So his later theory, the general
theory of relativity, worked out from 1907 to 1915, appeals to these more
complex motions, dealing with the laws of physics in different reference
frames, taking into account their relative velocity. A new view of space
and time was explained. The second theory destroyed the classical Newtonian concepts of absolute space and time.

Einstein realized that the requirement of constant velocity would
eventually be insufficient. The 1915 general theory removes this restric-
tion and deals with all forms of motion and also, for the first time, with
gravity. Einstein clearly demonstrated a new view of space and time,
since the Newtonian and Galilean rules did not apply to light. His postulates are described as the relativity of simultaneity, time dilation, and
length contraction. He proceeded to explain these concepts in various
ways.

The classical physics of the late nineteenth century comprised three
major and seemingly all-embracing fields: the mechanics of Newton; the
electromagnetic theory of Clerk Maxwell; and the thermodynamics of
Lazare James Carnot, James Joule, and others. To some, it seemed like
the end of physics, with little more remaining to be discovered.

That complacency was abruptly overthrown in 1905 when our
twenty-six-year-old technical expert third class in the Zurich Patent Office published papers on three groundbreaking topics. His papers on
Brownian motion (his doctoral dissertation) were decisive in persuading
most physicists that matter was indeed composed of atoms—as had been

widely speculated since the time of Democritus in fifth-century-B.C.E. Greece. Papers on the relativity of motion, attributing a fixed velocity for light independent of the velocity of the observer, implied the equivalence of mass and energy. Another paper explained the photoelectric effect in terms of Planck's recently discovered quantum, with the assumption that electromagnetic waves (light) could sometimes behave as particles. Einstein himself regarded only the last discovery as "very revolutionary," and it was for this that he received the Nobel Prize in 1922.

Quantum Mechanics: Its Mischief and Its Legacy

In 1923, Louis de Broglie proposed that particles could sometimes manifest as waves, thus completing Einstein's wave-particle symmetry. This paved the way for unification at a new level when quantum theory itself underwent a revolutionary upheaval with Erwin Schrödinger's discovery, in 1925, of the wave equation, now known by his name. At about the same time, Werner Heisenberg and Max Born in Germany and Paul A. M. Dirac in England independently discovered formulations that were eventually shown to be equivalent to Schrödinger's. Quantum mechanics was launched in the basic form known ever since.

Despite a major effort by Dirac and others to make a relativistic quantum mechanics, the theories failed at high energies. The solution required a third revolution. Then, in 1949, Richard Feynman and Julian Schwinger in the U.S. and Sin-Itiro Tomonaga in Japan discovered a new mathematical technique called "renormalization." The new theory, quantum electrodynamics, or QED, was fully relativistic and was able to make predictions of experimental results to unprecedented accuracies, providing a model for yet further developments. These were: quantum chromodynamics (QCD), which covers the strong force between quarks; quantum flavor dynamics, covering the weak force between leptons (electrons and neutrinos) and quarks; and the "standard model," which assimilates all of these theories.

The debate as to whether light consists of a stream of particles (Newton) or is a wave (seventeenth-century Dutch physicist Christian Huygens) goes back some three hundred years. The argument was settled when English physicist Thomas Young in the early 1800s devised a test that has become one of the classic demonstrations in physics—the double-slit experiment (which will be explained later in the book). This experiment defeated Newton's idea that light was but a series of particles, and confirmed that it is also waves. Maxwell and others, including Einstein, in some ways rescued Newton's particle theory of light from total defeat, thus confirming this conundrum mathematically.

30

How About Us? Are We Made of Waves, Particles, or Music?

Dr. John Bahcall from the Institute for Advanced Study at Princeton says: "We are all quantum fluctuations. That's the origin of all of us and of everything in the universe." It can be said that fluctuations are waves and, if so, that we are nothing but waves after all; we are mostly water. Hence men must be Neptunes or Poseidons and women mermaids. But before getting into the anatomical science of mermaids, I guess we better change the subject!

The impact on the world of Einstein's relativity theories, great as they were, is easily equaled by the second great discovery of the twentieth century, quantum physics. At the turn of the century, Max Planck, warm friend of Einstein, had an insight that would lead a quarter century later to the "quantum revolution" and a complete break with what we view as reality.

Planck opened the door by conceiving the idea that energy is not continuous but comes in packets of energy called "quanta." Though awarded the Nobel Prize in 1918 for this discovery, he never quite believed in his own discovery and kept looking for the flaw throughout the last years of his life. Modest, like Einstein, he wrote:

> My futile attempts to fit the elementary quantum . . . into the classical theory . . . cost me a great deal of effort. Many of my colleagues saw in this something bordering on a tragedy. But I feel differently. For the thorough enlightenment I thus received was all the more valuable. I now knew for a fact that the elementary quantum of action played a far more significant part in physics than I had originally been inclined to suspect.

Musical Einstein wrote about Planck:

> The longing to behold harmony is the source of the inexhaustible patience and perseverance with which Planck has devoted himself to the most general problems of our science, refusing to let himself be diverted to more grateful and more easily obtained ends. The state of mind which enables a man to do work of this kind is akin to that of the religious worshipper or the lover; the effort comes . . . straight from the heart. May the love of science continue to illumine his path and lead him to the solution of the most important problems in present-day physics.

Einstein felt that he owed a lot to Planck's discovery of quanta, for they had spent much time together discussing each other's ideas.

In 1925 Werner Heisenberg formulated the first description of quantum mechanics. He and other physicists tried unsuccessfully to test the principles and ideas against similar problems in classical physics. All was to no avail until Heisenberg discovered a completely new way that rendered classical mechanics useless. He came to realize that there is no way to accurately measure the momentum of a subatomic particle unless you are willing to be very uncertain about its position, nor can you pinpoint its position without being uncertain of its momentum. This "uncertainty principle" overturned all previous "common sense" knowledge of the subatomic world, just as had Einstein's relativity theories on the cosmological world.

It is highly relevant to point out that Heisenberg, one of the giants of the physics world, was accomplished on the piano and the violin. He made frequent allusions to music in his scientific discussions; he understood the profundity of music in the universe.

Who could not be astounded on first learning of the two experiments: the "double slit" experiment and much later the realization of the uncertainty principle that has fed science fiction writers not to mention controversy and further research by serious physicists ever since.

Science Fiction, Where Are You?

The most amazing phenomenon of all, in the opinion of this musician/amateur scientist, has been realized again and again in experiments by hundreds of physicists. This astonishingly true story occurs when a single particle is shot at great speed through the tunnel of an atom-smashing machine, so constructed that the unfortunate particle hits an obstacle, such as a mirror, and is shattered into two particles. These twin particles now continue at great speeds in different directions, let us say like the *left* and *right* arrows on your computer. They could be traveling yards, a mile, or trillions of miles, and when one twin is diverted (perhaps by a mirror or other obstacle), the other, by this time a trillion miles away from its sibling particle, *simultaneously* goes in the opposite direction (perhaps like the *up* and *down* arrows on your computer). When I first heard of these experiments, I was overwhelmed. Can it really be true that these cosmic racing twins can be deflected so that one continues in one direction and the other goes in the opposite direction; then, traveling away from each other at enormous speeds, one is again deflected, say at right angles from its path, and the other particle, merrily on its way, no matter how far away (a mile or a trillion miles), *simultaneously* changes its course, also at right angles but directly opposite from its twin!

Think about it.

Why is the implication so mind-shattering? Can one then extrapolate that everything in the universe is in constant communication with everything else? Can this communication exceed the speed of light, which it must if the communication is instantaneous!? Can the phenomenon be true on a human or cosmic scale? Or, disappointingly, is it limited to the very small? If the macroworld indeed differs only in scale from the microworld, then this experiment, demonstrated over and over, indicates the unity and oneness of all phenomena! It could easily be interpreted to mean that there is some mysterious, simultaneous communication between all particles in the universe. Which opens a question: Are there no actions completely separate from any other action? I suspect so.

We certainly have a lot to learn about the universe, both large and small. What is real and what is not? And how does this apply to our world?

A classic example, used to describe the vagaries of what is real and what we *perceive* as real, is a painting by René Magritte. Under the simple painting of an ordinary pipe is a message, written in script (which is almost as bad as my handwriting), *Ceci n'est pas une pipe* ("This is not a pipe"). Magritte is telling us that this is not a real pipe, only a painting, so please don't smoke my painting! Magritte obviously had a more serious message in mind, since he returned to this theme many years later in another, similar painting. As Timothy Ferris, author of many excellent books on astronomy and physics, writes in his book *The Mind's Sky*: "The point is to challenge the belief that everything outside the frame is real." Ferris comments:

> The enemy of surrealists like Magritte, and of artists generally, is naïve realism—the dogged assumption that the human sensory apparatus accurately records the one and only real world, of which the human brain can make but one accurate model. To the naïve realist, every view that does not fit the official model is dismissed as imagi-

nary (for those who "know" that they err when they entertain contradictory ideas) or insane (for those who don't).

None of this would have emerged without quantum physics, which is
confined to the very small; only recently have physicists managed to
write a quantum mechanical description of something as large as a single
molecule, and molecules are a billion times smaller than human beings.
But this does not mean we can ignore the implications of quantum-
observer-dependency for our world. Big things are made of small things;
we too are made of atoms, and quantum effects influence our macroscopic world as well as the unimaginably tiny.

Travels in Space

To confuse the matter further, consider the following classic example, an
Einstein thought experiment: Let's say you are standing on a platform at
a train station while a train is passing by. Lightning bolts strike, one in
the front end and one in the back end of the train as it passes. The lightning leaves char marks on the train and in two corresponding points on
the platform. You notice that the light from the lightning reaches you at
the same distance. Pacing off the distance between your point of observation and each of the char marks on the platform, you see that the two
distances are exactly equal and that the speed at which light travels to
you is the same. Thus, you assume that the two bolts occurred at the
same time.

Now let us look at the same two bolts from the perspective of an observer on the moving train. This person's reference frame is at the center
of the train, and because he is moving with the train, he is approaching
the light signal, which travels toward him from the front end of the train.
At the same time, he is moving away from the light signal at the back of
the train; thus he sees the front-end light signal a very brief instant before
the back-end signal. He also finds that the two distances are equal. He
concludes he is equidistant from the two events. He concludes that the

light from the front reached it earlier than the light from the rear. Thus the two events are simultaneous for you at the station but are not to the observer on the train.

Einstein's response is that both you and the train traveler are right. The simultaneity of two events is not an absolute notion for both observers. The amount of time delay between two events is different for each person in differing reference frames. Thus, simultaneity is a relative concept, not an absolute one.

Travels in Time

If a rocket passes the earth at a high speed, then an observer on earth will see the clock on the rocket as moving slower than a clock on earth. At the same time, the pilot will see the earth clock going slower than his on-board clock. Each will see the other's clock slowed down. The observer will also measure the length of the rocket at high speed to be shorter than it is at rest. The invariant speed of light makes all this credible and logical. Just as the simultaneity of the events is not absolute but is dependent on the reference frame, so too is the interval of time of the event also dependent on the frame of reference.

These concepts, explained relatively simply here, when extrapolated into higher mathematics for the cosmos in its vast dimension and the microworld of particle physics in its incredibly shrunken dimension, have enabled later scientists to utilize atom smashers and giant accelerators, which eventually resulted in atom bombs and interplanetary rockets. It is also the stuff of science fiction where, say, one person takes off from earth in a high-speed rocket and leaves his friend back on earth. If the rocket travels at nearly the speed of light and makes a round trip to a star in the course of, say, ten years, then the earth-bound friend would have aged ten years. Meanwhile, the traveler would have aged perhaps less than one year.

But instead of suggesting that everything is relative, Einstein shows a clear order and seeming permanence beneath the appearance of change. Einstein's theory is about invariance, harmony, and light.

The Lord saith . . .

> *"And God said, 'Let there be light,' and there was light. And God saw the light, that it was good: and God divided the light from the darkness."*
> —*Genesis* 1:3–4

Though Einstein lifted the concept of light to entirely new levels, much speculation about light had preceded his ruminations. As he gracefully admitted, he had many shoulders to stand upon. The first chapter in the Old Testament, as in the statement above, uses light to describe the beginning of the world. Light has been a metaphor ever since biblical times—light the way, en-lightened, shed a little light on the subject, lighten up, etc. From time immemorial, the concept of light has engaged shepherds, artists, philosophers, the religious, and scientists. Many scientists were concerned with light as the very core of human existence. Aristotle thought that light is not a substance but an activity, though he was mostly interested in vision and how the human eye perceived the world.

Scientific work on the subject abounds, especially in the nineteenth and twentieth centuries. Sir Isaac Newton held that light was composed of particles that he called corpuscles. Heinrich Hertz, who had previously discovered radio waves, discovered the photoelectric effect in 1887. And, of course, light is at the core of Einstein's theories.

What Time Is It in London?

Relativity has overturned our most basic sense, our experience of space and time. Our clocks, largely based on natural light (day versus night), tell us what time we should be at work, eat, and go to bed, and it is an established fact that the clock will tell us the time of day. So light is intimately tied up with time and space. By the same token, we know the distance between one body and another, ourselves, for example, and a being in another state or country on earth.

The word "geometry" indicates that the concept of space is connected with the earth, or "geo," which is a solid and ever present body of reference. In the geometry of the Greeks, the positions of a body in space was considered constant. In Descartes's language, all of Euclidean geometry is axiomatically founded upon (1) Two points of a rigid body determine a segment, and (2) the distance between two points can be measured permanently, provided neither body moves. This confirms our *sense* experience. However, even in our very limited sense experience, time does not seem as rigid as space; we all experience, for example, the differences in time when we cross time zones in our travels.

Einstein's general relativity revealed that space becomes curved as a result of gravity and that bodies change in size according to gravity. Indeed, gravity takes its toll in all of us throughout our lives. Some of us lose inches in height as we age. When drawn to the logical conclusions, we find ourselves in an incredible morass of mathematical complexity in order to realize "objective" space and time. It boggles our subjective minds and sensory experiences. Classical mechanics are totally inadequate to describe macro or micro phenomena, and therefore we have to arrive at conclusions by way of mathematical equations and similar esoterica.

A Simple Musical "Theory of Everything" (TOE)

Yet most physicists believe that nature has an ultimate simplicity. The fundamental laws of physics are surprisingly simple and comprehensible. These laws can be divided into three parts:

1. Particle physics, which concerns itself with the smallest, most microscopic particles in the world,

2. Gravity, from Newton to Einstein and the general theory of relativity, and

3. Cosmology, which deals with the apparent birth or death of the universe.

These three sciences have been converging for some years, though gravity remains stubbornly elusive. Physicists are hoping to find one ultimate theory that will explain everything. Superstring theory, many of its proponents believe, holds the answer to the explanation of all phenomena. That's why it's called the "theory of everything," or TOE.

Superstring theory postulates strings, smaller than any known particle, that vibrate and undulate according to wave motions. These strings are envisioned as the ultimate, most basic elements. Physicists frequently describe the vibrations and waves of musical strings to explain phenomena in this theory. Our musical theory takes the next mini-step by postulating that music is the very substance of the universe; in other words, music could supply a theory of everything. Simple?

The basic elements of music, like those of physics, are simple. A small child can learn time signatures, the up and down of the scale, and the loud and soft of dynamics. Still the mysteries remain: How can one explain, for example, why the harmonic series exists throughout nature and the universe; or why the periodic table of the chemical elements is based on integers (whole numbers); or why the electric charge of the nucleus is an integer? Why are they not, say, one and five-eighths or seven-sixteenths or some other weird number? How come they are integers? It's as if God were a fourth-grade arithmetic teacher trying to simplify his teaching to a backward species. I suspect that my wife's "Messianic," fundamentalist cousin would like that explanation.

31

Quantum Mysticism

*"Books on physics are full of complicated
mathematical formulae. But thought and ideas, not
formulae, are the beginning of every physical theory."*
　　　　　—NEILS BOHR, Nobel Laureate

I n the above quote, try replacing the words "physics"
and "physical" with "music" and "musical." Musicians
would generally agree that music is also a science, and
most scientists would probably agree that science is also
an art. The common belief, however, has been that science or the scientific method abjures myth and faith.
That long-held belief has been challenged in our post-
modern era.

Quantum theory has put the spiritual and the mystical front and center. Myth and faith are not alien to musicians, whereas scientists have been generally careful to
safely enclose their visions and even their mathematical
certainties in the word "theory." They know very well
that what were considered certainties not many years
ago have since been disproved. Former axioms have
been shattered and are now dependent on contextual

frames and grids, leaving a blank canvas upon which the painter, composer, and scientist paints his vision. Do you and I have that same visionary blank canvas available?

The Window Is Foggy

Postmodernists, characterized by extreme social fragmentation, take refuge in perversions of quantum theory and even in the 5,000-year-old Vedic sages as well as contemporary Hindu gurus, suggesting that the universe is simply the result of our consciousness. We're imagining the whole thing! But if you are skeptical, it is not so easy to dismiss the scientific major leaguer Neils Bohr. On his personal coat of arms, he utilized the Taoist yin/yang symbol, which suggests that if the earth is our mother, the sun is our father.

Musicians took up symbolism early and enthusiastically. Our friend Beethoven was interested in Greek philosophy and mythology, and he incorporated the Promethean legend in his "Eroica" Symphony and in *The Creatures of Prometheus*. The Hindu philosophy makes an impact up to this day. When the Age of Reason brought Cartesian science and reason to the fore, music and science were looked on through a different window, which fogged up after quantum mechanics took hold. Bohr and Heisenberg struggled with the paradoxes that quantum physics posed, until they finally came to the conclusion that there can be two aspects of the problem, both true, and that the two, taken together, offered science a more complete perspective on the atomic world. Bohr showed that the components could be combined into a greater harmony, each complementing the other.

The art of music arrived at such a paradox some centuries earlier, when it was no longer necessary, according to the Church, to limit music to singing and playing in unison. The wonder of counterpoint, two separate lines of music running concurrently "against" each other, became fashionable. What marvelous musical counterpoint by Bach and others we might have missed if the ban on polyphony had not been lifted!

Later, when harmony (three or more lines of music played simultaneously) came on board, the effects were even richer. The processes in both
physics and music can be described in terms of the Hegelian and Marx-
ist "thesis, antithesis, and synthesis."

The quantum discoveries in science have changed the way we have
viewed the world for thousands of years. As exhilarating as these
changes are, they have also provoked acrimonious controversy between
secular science and the religious worlds, extending out to sociology, philosophy, and politics. It has provided ammunition for the most diversi-
fied disciplines and opinions, putting extreme opposites in bed together,
from religious mystics to serious scientists.

Authors added their voices to the battle. In *Late Night Thoughts on
Listening to Mahler's Ninth Symphony,* biologist Lewis Thomas writes:
"Any effort to insert mysticism into quantum mechanics or to get mysticism out of it . . . is not for me. . . . It is not true to say that every man has a
right to his own opinion. I do not have the right to an opinion about causality, black holes or other universes. . . . I concede it is none of my business."

Novelist Annie Dillard adds: "Nothing is more typical of modernist
fiction than its shattering of narrative line. . . . The use of narrative col-
lage is particularly adapted to twentieth-century treatments of time and
space." She believes that there is meaning in the world, cause and effect,
concluding that lyric poets are best equipped to find it. Thomas and Dillard are only two of the many contemporaries involved in the controversy centering around the three worldviews that have gained relevance
during the last century, after quantum mechanics came onstage:

1. Causality
2. Acausality, also called postmodernism
3. Process thinking, or dialectical materialism

At first these concepts were exclusively in the vocabulary of philosophers, academics, scientists, and a few historians. Today these terms
have invaded the everyday language of news dailies and popular and serious magazines. Let's take a moment to see what they are talking about.

Causality was the basis for scientific thinking in the couple of centuries previous to Einstein. Arising out of Newtonian and Cartesian thinking, it explained the world very logically. It was simple and useful: A causes, B, which causes C, which causes D, etc. This makes sense for the cause-and-effect world we experience in our daily lives. Our three-dimensional world gave us the tools for the industrial revolution, the Age of Enlightenment, and a modern technology that promised to liberate mankind and make life easier—with widgets, washing machines, atlases, and automobiles. We can observe its considerable accomplishments all about us. This very practical approach still serves a most useful purpose in our "real" world. But it suddenly had nowhere to turn when Einstein's theories began pulling the ground from under its feet, aided and abetted by the resounding emergence of quantum mechanics.

We now had to view the vast macro-universe and incredibly tiny micro-universe in much broader terms and dimensions. What was previously only the stuff of science fiction suddenly became either real or realizable, proven by experiment after experiment. Meanwhile, the combination of Newtonian technology, Einsteinian theory, and quantum mechanics began producing a nuclear Frankenstein that threatened to destroy us all.

Many of the followers of the second worldview, acausality, were impervious to this danger, for they were metaphorically straightening the deck chairs as the great ship *Titanic* was sinking into the sea. In acausality, or postmodernism, sometimes labeled "deconstructionism" or "reductionism," nothing causes anything else. Quantum mechanics, Werner Heisenberg's uncertainty theory in particular, fed into this worldview that nothing is as it seems—though neither he nor Niels Bohr would, I think, endorse the cloudy postmodernism espoused by French writers Jacques Derrida and Michel Foucault and abetted by some American intellectuals and right-wing political activists. For these, the world is a game of chance. Unlike the pundits, philosophers, and comfortably funded scholars, a poor man trying to feed his family has no problem defining reality. For him reality is simple: food and shelter. The reality is equally simple for a human rights prisoner, unjustly incarcerated for twenty years; or a Vietnam veteran, sick with Agent Orange.

Are we all pawns to nature's dice? Or does the Great Designer in the Heavens fatefully detail the script in advance? The ramifications of quantum mechanics and the uncertainty principle caused all kinds of problems and controversy among scientists as well as between the secular and religious worlds.

Uncertainty theory elicited Einstein's famous quote, "God doesn't play dice with the universe." He held on to causality as long as he could, but late in life he had to reluctantly surrender to at least part of quantum theories, albeit with reservations. His unwillingness to agree with some of the conclusions arising out of quantum mechanics remained with him until his death. He was not about to give in to the mysticism that many derived out of quantum theory. Einstein's head was in the skies, but his feet never left the ground.

Einstein wanted to talk with the famous comedian Sid Caesar. Unfortunately, Einstein died a week before the meeting. Physicist Robert Oppenheimer told Caesar that Einstein had wanted to talk about "the human equation." My guess is that he grudgingly accepted the "uncertainty principle" for the microworld but believed and acted in the macroworld and the everyday world according to causality. In other words, events occurred with a causal, testable connection. Yet he surely felt there is room for spontaneity and choice. With enough numbers (of locusts, ants, or people), group behavior, such as in a church mass or a political demonstration, becomes more predictable. Above all, Einstein was a humanist, caring about real people in a real world, "real" as defined by common usage.

F. David Peat, in *Synchronicity: The Bridge Between Matter and Mind,* stated that "logic depends upon a whole context, which itself depends upon a series of contexts. The whole meaning belongs to the entire system. All things in nature are related and unrelated, different in some ways and similar in others. These do not exist in the objects themselves as in the act of observation—which always takes place within some particular context. This context may be sensitive to a wider context which is itself embedded in an ever-changing world."

The deconstructionists were busily pulling apart and examining unconnected details. Each detail was equal to the other and had no relation to the other. This tendency provoked the quotations at the beginning of this chapter, and the following one from Joan Peyser, author of books on modern music, as she weighs in on deconstructionism.

"Analysis annihilates mystery," says Peyser, as she expresses dismay at the trend in contemporary music that she attributes to the theories of Freud, Darwin, and Einstein. Why she includes Einstein I'm not sure, except, perhaps, that he dismantled the universe as a whole and explained how the parts work, though this was all in the context of each part relating to the whole. I suspect that she would find common ground with Einstein, who fought all his life against what he perceived as the obscurantism arising out of some interpretations of quantum mechanics.

Lewis Thomas, in addition to being a loved and respected scientist, was also an ardent music lover and a practical man. In the same book quoted earlier, he says, "Surely music (along with ordinary language) is as profound a problem for human biology as can be thought of, and I would like to see something done about it." He tells of the German government appointing a large advisory committee to work on the question of what the Max Planck Institute should be taking on as its scientific mission. "The committee emerged with the recommendation that the new Max Planck Institute should be dedicated to the problem of music—what music is, why it is indispensable for human existence, what music really means." Then, sarcastically, "The government, in its wisdom, turned down the idea, muttering something in administrative language about relevance."

Scientists, artists, and poets do look more deeply into things. In general, their work leads them to reject acausality. The causal links are often built of complexities. They are not only linear and confined to three dimensions, but they must also be considered in many dimensions. These dimensions are not things in themselves; they too are subject to constant interweaving, dynamics, and change. The uncertainty principle holds here as well. Yet, none of this can be considered as separate from the conditions and historical background from which they emerged and in which they function.

It is useful to revert to our primitive, prescientific man. He had a busy time looking for food and protecting himself and his family. As his progeny grew up, they were recruited for the task of foraging for food and keeping alert for any dangers nearby. Soon that was not sufficient, so humans teamed up with other humans and families to defend one another and to help in the hunt. In the agricultural ages that came next, they divided into tribes, which would compete with one another for a particularly good piece of ground or a good watering facility. Thus began the very concept of competition. Along with the beginnings of competition came the first seeds of "civilization" and strife.

In competing, a tribe sought the most advanced tools or weapons available. The science of weaponry at the time (stones, bows and arrows, etc.) became useful for either digging or destroying. This science of building bigger and better weapons has a long history. The scientist who could "invent" these usually had a place of honor, although scribes and songsters—those who could pass on the tales of previous exploits and ideas—were equally honored. The importance of both music and science can be seen in all periods as far back as recorded history.

The causal, determining forces are the way people make their livings and the way they relate to other beings. Science, music, and life continually accumulate data to modify or discard old theories and formulate new ones. In fact, this writer doesn't believe that there is any one answer, now or ever, that will explain all phenomena. There is no end, nor was there a beginning.

The deconstructionists deny process, predictability, and responsibility. They eschew relationships between phenomena and thus falsify the nature of the universe. Right and wrong, truth and falsehood fall by the wayside. All are equal by their lights. Anything can happen in a world that is simply a game of chance, where we are all pawns to nature's dice. Conclusion? It doesn't matter what we do. Any behavior goes, however vile, evil, or dishonest, no matter the consequences.

However, the debate between causality and acausality is moot. Both are right and yet both missed the boat. Neither is sufficient by itself. The answer lies in the interaction between the two and an inescapable third entity, reality.

32

What Is Real?

Quantum mechanics has led many scientists to re-define reality. Fiction writers, science and other-wise, have jumped right in with their own concoctions. One of the many dictionary definitions of reality is "the state of being true to life." We can ask: "Whose life?" Reality is experienced differently by any two people, any two groups, and any two countries. So do we have any firm ground to stand on? Where is beauty, numbers, symmetry, the arts? If reality is the realization of true art, let's say a Bach cantata or a Mozart symphony, what is, say, one artist's conception of wrapping up a city, or a Bronx youngster's graffiti on subways? What about the reality of refugees fleeing anywhere to avoid torture, starvation, and death?

What was Bach's or Mozart's sense of reality? Both spin out their wonderful detail, so perfectly fitted together that the end product sounds "just right." Beethoven wrote four overtures to his opera *Leonore* until he thought he got it right, though to our ears, each has its own merits, or should I say magnificence. But like a painting, we may not understand the artist's reality, but we know what we like, we know reality when we see it—or think we do!

Scientists are still trying to "get it right." The incredible amount of detail that has been amassed in the last century, and that is still growing at a geometrical pace, has yet to be put together in such a way that it yields a provable theory of everything, a cosmic reality, a grand symphony.

One broad approach to reality is "process thinking," important because, for this writer at least, it answers virtually every previous question and perhaps any conceivable question. Process thinking refers to the fact that the universe is in a continual process—is always changing—in all life, politics, the heavens, and the atom. Nothing is static, nothing is rigid. All is in flux. The process approach encompasses and synthesizes previously described approaches. It may yet yield the "unfinished symphony" yearning to be written. It helps predict the stock market, evolution, revolution, the cycle of life and death. It takes on the task of joining together the huge amount of data and experience of visionary leaders, scholars, and people in general in an effort to find, among other things, the "theory of everything." It might even clue us in to a new and greater music.

Reality arises out of the dynamic, ever changing interaction of all phenomena, a living theory of thesis and antithesis. It is a reality that acknowledges the obsolescence of racism, the cutting up of our globe into artificial, long-outdated nation-states, religions, races, and other factions and fragments.

Can the process-thinking mode of global operation bring us at last to Beethoven's "Ode to Joy" declaration of the oneness of all humankind? Only if the theory is accompanied by practice—the unending struggle for a better world. Both theory and practice are necessary: Theory without practice becomes little more than high-sounding philosophical banter, while action without scientific theory behind it has led, time after time, to dead ends and destruction.

Unless something is done to revive the reality of what it is to be a human being, to halt the dismantlement of moral and ethical values and of the entire earth environment, to begin the rebuilding of our earth home, and to get quickly to these tasks, nature's reality will continue to punish us all, guilty and innocent, rich and poor alike. But we need not be willing participants in our own demise.

An earlier chapter relates the argument between my fellow students on objectivity versus subjectivity in music. You may recall that I felt the answer was that both are right. By separating one from the other, both entertain a false perception of reality. They need each other to find the total truth, from the approach of learning, to playing an instrument, to music itself.

Musical Space-Time: Perspective, Spontaneity, and Freedom

Time and space are prime examples of the Marxian/Hegelian concept of thesis, antithesis, synthesis, or reality. As early as 1877 the composer Richard Wagner used the words "here time becomes space." He certainly could not have known that these words would take on literal meaning some thirty years later in Einstein's theories. Implicitly and sometimes explicitly, Wagner, Beethoven, and other composers seem to have intuited these concepts, as reflected in their music.

We are accustomed to viewing the three dimensions of physical space—length, width, and depth. Scientists add a fourth dimension, time. The depth dimension, for us, is constructed by the mind from our binocular vision. Much of what we see depends upon mental interpretations, which are constructed from the patterns of light, dark, and color that fill our vision. The light that strikes the retina of the eye triggers nerve impulses that are carried from the nerves to the brain, and then interpreted in the optical centers of the brain as those objects in the space around us.

For the Greeks, perceptions, feelings, and thoughts were forms of motion in space, and there was little demarcation between the mental and the physical. So after thousands of years of man developing his view of the universe, Einstein took a final leap and reached the top of the mountain, where one could see the universe in new ways.

You Can't Do That!

Since frames of reference play such an important role in physics and in all life, I would like to recount an experience I had when I was giving a lecture at a German music conservatory some years ago. After the talk, a group of students asked if they could accompany me back to my hotel to ask questions related to my talk. Walking along, we arrived at a wide boulevard, where we faced a red light. We looked up and down the long avenue, and there wasn't a car in sight. Being a New Yorker, I started to walk across the avenue. After a couple of steps, I realized that the students were not following.

Instead, the students looked in horror, saying, "You can't go against the red light!" Retracing my two steps to the sidewalk, I said: "Come with me, it may be one of the most important music lessons you ever had." Against their better judgment, they reluctantly crossed with me, quite safely, of course. When we reached the other side, I was amused by their exuberant reaction, as if they had been suddenly freed from shackles in a prison. One student said to me, "Yes, we Germans approach everything, even our holidays, in a serious and structured manner, with little flexibility. Everything is planned and followed through carefully. We have a German expression to describe ourselves, *tierische Ernst,* which means 'animal-like earnestness.'"

Music, which creates order out of disorder, clearly requires structure. However, it must also have spontaneity and freedom within that structure. Music, like everything in our lives, must be looked at in its dynamic, ever changing context—its frame of reference. This includes absolutely everything—you name it: music, politics, science, society, and crossing the street.

Auras (with halos
at an extra price!)

Science affirms that "auras," or energy fields, surround all of us. People, just as they are, without benefit of musical instruments or song, could be said to *be* music. When we go into a room filled with people we have never met, we have trouble catching their names because we are too busy "reading" the music that emanates from them. This exercise can become quite a game. Try it.

Unfortunately, the host or hostess too often has background music playing, especially when they know a musician is coming. Being careful to avoid insulting the host, I ask for the music to be turned down or off. Most of the time I prefer to "hear" the music of individuals and of the group dynamic. It tells so very much about people if we only allow our inborn human sensitivity to assert itself.

All of us could hear this music if we would only open ourselves to the awareness of its pervasiveness, even in a supposedly silent forest in the dead of winter. How sad it is to witness the self-induced narcosis of young people who turn up their boom boxes to ear-shattering decibels every- where they go—in their cars, in their homes, and on the street. What tragic music in themselves and in their lives are they covering up? How different it is to hear a small classical chamber group, or an acoustic folk group. Once, I had the pleasure of hearing a handful of Celtic folk musi- cians in a pub in Asheville, North Carolina, intimately playing and singing for and with each other. The beauty and integrity of their music gradually quieted the cacophony at the bar. People slowly drifted closer to hear these sensitive musicians. No big bang was necessary to get their attention.

The Musical "Big Bang"

On April 29, 2001, astronomers monitoring two detectors in Antarctica found traces of colossal waves, much like sound waves, emanating from space. The minute patterns in a glow from primordial gases may have ignited the big bang some 14 billion years ago. The astronomers speculated that the patterns were probably created by energy fluctuations at the quantum scale when the universe was a tiny fraction of a second old, when it was smaller than a tennis ball. The detailed observations detected the fluctuations from the glow of the hot gases, or "cosmic microwave background radiation," that carried an imprint of those waves to the detectors.

Dr. John Carlstrom, a University of Chicago astrophysicist who leads the team operating the Degree Angular Scale Interferometer (DASI, pronounced "daisy"), a microwave detector at a South Pole research station operated by the National Science Foundation, exults, "We see the structure of the universe in its infancy."

"We are living in the most exciting time ever," adds Dr. Michael Turner, a cosmologist at the University of Chicago. Others involved in the announced discovery included California Institute of Technology astronomers and the so-called Boomerang team, made up of astronomers from Italy, Canada, and Britain. Dr. John Ruhl of the University of California at Santa Barbara presented results of the Boomerang Team at a meeting of the American Physical Society. Another team, Maxima, with astronomers at the University of Minnesota and the University of California at Berkeley, reported less distinct observations of the wave patterns.

The Music

An article by James Glanz, reporter for *The New York Times,* expands on the musical connection: "The leading theory of how the universe could have exploded out of the primordial nothingness, known as the theory of inflation, predicts that the quantum fluctuations should have rattled the universe in such a way that it resonated like a vast organ pipe, with a main tone, overtones and a series of harmonics. Imagine colossal Super Strings playing in a cosmic cathedral!"

Last year, the Boomerang team detected the main tone but found no clear evidence for the overtones, raising the possibility that the inflation theory could be wrong. Since much of the information about the fluctuations, such as their relative intensity and overall spectrum, would reside in the characteristics of the overtones, those results raised the prospect that few remnants of the initial spark might be found. But recently the three teams announced that they had seen two of the overtones for the first time. In musical terms, they observed the first two harmonics above the main tone. "We do see two more bumps and wiggles out there," Dr. Ruhl said. "We can move to the question: What do these bumps and wiggles tell us?"

Dr. Max Tegmark, a cosmologist at the University of Pennsylvania, said that while the new results were still far from absolutely proving the inflation theory, their agreement with the theory was uncanny and would cast doubt on alternative models. "It's even scary that things agree this well," he said. "This is a very bad day for the competition."

The Cosmos Hums Its Tune

Other scientists, including Dr. Andrew Lange of Caltech, a leader of the Boomerang group, said the results strikingly showed cosmologists the composition and behavior of the universe in the first few hundred thousand years of its life. It was then that the sound waves were humming through the young cosmos. Astronomers believe that the microwave background radiation was emitted as the universe cooled below a critical temperature when it was about 400,000 years old.

"We've really been waiting for the other shoe to drop," Dr. Lange said in reference to the lengthy search for the overtones. "What we're confirming for the first time is a very generic prediction of modern cosmology."

Although astronomers said that much more detailed observations, including the discovery of more overtones, would be required to define the quantum fluctuations and to verify inflation, the results are likely to be seen as major victories for two scientists in particular. Dr. Alan Guth of the Massachusetts Institute of Technology, developed the germ of the inflation model in 1980, a theory he called "the ultimate free lunch" because it shows how the entire universe could have exploded out of nothing and impressed the quantum fluctuations on the cosmos. The results also provide major support for ideas closely associated with Dr. David Schramm, a Chicago cosmologist who died in a plane crash in 1997. Dr. Schramm and his colleagues worked out a theory, unrelated to inflation, using trace elements created in the big bang to gauge the amount of ordinary matter in the universe. Those values agree closely with the amounts deduced from the intensity of the sound-wave overtones affected by the sloshing of matter in the peaks and troughs of the sound waves.

33

Extraterrestrial Music

If one assumes, as most scientists do today, that wavelengths of light and wavelengths of all matter obey the laws of mathematics and the laws of the universe, then waves are a universal language, direct from nature.

We know that music can be explained by mathematics just as mathematics can be used to describe music; thus it may not be so far-fetched to say that music is, at the least, a guide to a theory of everything. Indeed, music could be seen as the very stuff of the universe. If one wished to send a message to other worlds in outer space, the one message that has a possibility of being understood would be music. On this cosmological scale, and since we are postulating that the universe is music and that music expresses and explains the universe, then we can take the next logical step, that music could hold the key to a T.O.E.

Professor Witten has argued that the sheer number of heavenly bodies in the universe makes it very likely that there are other life-forms out there that are at least as advanced and probably more advanced than we are. Hence, if we were to communicate with these life-forms or they with us, the means of communication would probably be via radio waves or music. Composers take notice!

Most people today do keep an open mind on the question of extraterrestrial life. Since wavelengths of light and wavelengths of all matter obey the laws of mathematics and of the universe, waves are the truly universal language. If we were to communicate with extraterrestrial life, music would indeed seem to be the logical form to use. Since music expresses and explains the universe, music brings us closer to understanding what (and who) goes on in the cosmos.

An extraterrestrial language therefore could be music or mathematics. A Mozart waltz gives eleven possible variations for fourteen of the sixteen bars of music, with two additional options for the performance of one of the other two bars. A mathematician tells me that this gives 2 times 11 to the power of 14 possible waltzes. Enough to keep millions of outer space couples dancing for billions of years!

The Absurd Universe

The universe seems to be as symmetrical as Beethoven's Fifth Symphony. The rules are remarkably similar for both the micro and the macro worlds. The question that physicists have been facing for many years is how to demonstrate that symmetry in the four forces (i.e., electromagnetism, the strong force, the weak force, and gravity). The first three have combined or at least have been related mathematically, but gravity has eluded any substantial relation to the other forces. Once gravity is combined with the other forces, physicists will have found the key to everything—the "grand unified theory."

For at least a generation, and despite regular excited revelations in the press, physics has made little progress—with the stunning exception of string theory. Mathematically, string theory shows signs of being able to combine quantum mechanics, Einstein's general theory of relativity, *and gravity.*

Though gravity is the weakest force by far, at least in our current understanding, there is so much of it that it bends the fabric of space-time, as has been demonstrated in the macroworld of stars and planets. It also exerts its force upon the very tiny existences within the atom. Fortunately, it also keeps you and me glued to the rapidly revolving earth, preventing us from being thrown off into space.

As discussed earlier, the strings of superstring theory vibrate in at least ten dimensions, with the extra six dimensions too small for us to see even with our most advanced instruments. These are curled up like a garden hose. From a distance, the hose looks like a one-dimensional object, but as we move closer, other dimensions appear, such as the circumference of the hose. If there were very small beings living in that garden-hose universe, they would be unaware of the extra dimensions. If two of these small creatures met, exploring just the length of this hose, one would be unable to pass around the other. But, if the circular dimension were inflated, the beings would have a new direction in which to move. String theory has not yet been developed to the point where it can make testable predictions about these dimensions. Here a musician might have the advantage.

Living in a world of musical abstraction, I can dream up all kinds of weird thought experiments. One of these I have been playing with is related to scale. If one takes a string and elongates it to cosmic scale, that string would be subject (like everything else) to curved space-time. The vibrating string, whether from a violin, or of quantum size, will eventually curve back on itself, like the superstring loops previously described. One can speculate that this vibrating string and its fellows could create a kind of "heavenly music" imagined by Kepler and other scientists before and after. The special wave and frequency patterns that we call music and that elicit universal responses from all life (not just from human beings but also from animals and plants) extends throughout the cosmos. Only humans have adopted the questionable benefit of linear thinking and memory, yet they still react to music.

Music has been recognized as the most abstract of the arts. While painting, sculpture, theater, film, and dance can evoke images of real-life situations, music creates a metaphor in listeners' ears, minds, and bodies. This metaphor conjures a plethora of memories—individual, societal, and, as Carl Jung would argue, of the "universal mind." True, the waves and patterns of music can be mathematically explained, but its residences in mind and body, both individual and societal, are less easily calibrated. And how about that mysterious macroworld, way out there?

Never detected directly, most of the cosmos seems to be made of so-called dark matter and dark energy lurking in space. Dr. Turner of Chicago has said that skeptics might well term that picture "the absurd universe" or "the preposterous universe." Sir Martin Rees, astrophysicist at Cambridge University, has said that scientists were left with the question of whether fundamental physical laws would someday explain that strange mixture of ingredients, or whether the precise amounts were a sort of accident of how the universe came into being—something like snowflakes, each of which has a hexagonal symmetry but carries a pattern that is otherwise unique. "It may well turn out that the underlying laws do not give us these numbers, any more than they give the detailed pattern of a snowflake."

Antigravity, Dark Matter, and Wormholes

> "To call Divinity music is to speak to the sound that all beings are making: All atoms are busy vibrating and making music in the universe."
> —HILDEGARD VON BINGEN (1098–1179)

Allusions to space travel, wormholes, and extraterrestrials are pesky and mysterious. Well beyond science fiction, these have invaded genuine science journals and a broad spectrum of mainstream media. One need only examine the testimony of dozens of governmental and military witnesses as well as people from all walks of life—doctors to dock workers, cooks to CEOs—who claim to have seen unidentified flying objects (UFOs) and/or been visited by extraterrestrials (ETs). Without free energy, how could spaceships cross the vast distances between galaxies? If any substantial part of the testimonies and research are authentic, then a revolutionary breakthrough is near. Scientific research into free energy from space, matter, and even time has occupied the serious efforts of scientists and philosophers for many centuries. Psychiatrist Dr. John Mack of Harvard gave credence to his many patients who are sure that they have seen UFOs and experienced the company of ETs.

Philosophers have been advancing theories on the relativity of motion and the dependent character of abstract time since as far back as Lucretius, who said: "Time exists, not by itself but simply from the things that happen." The time question remained unanswered for ages through discovery after discovery. From the mystery of time we have concocted all kinds of science fiction stories and scientific theories that sound like science fiction. In *Cosmos,* Carl Sagan relates a saga of alien beings sending a radio message to earth with details for constructing a machine capable of very fast travel to the center of the galaxy, to be achieved by traveling through a "wormhole in space." The fantasy was given a certain amount of respectability by John Wheeler, of "black hole" fame. Einstein had anticipated the geometry involved in wormholes, also known as the Einstein-Rosen Bridge.

Physicist Kip Thorne and his team at Caltech set out on a program of research to investigate wormhole possibilities and antigravity (a field that repels rather than attracts). Quantum physics provided some mathematical studies indicating that certain quantum states can produce antigravity. More recently, Richard Gott of Princeton University employed "cosmic strings," speculating that these threadlike "tubes" might supply the time "tunnels."

Speaking of time travel, conductors insist that the musicians arrive at rehearsals and concerts "on time" (actually they must arrive ahead of time to warm up their instruments, lips, fingers, etc., to be ready to play at the appointed time). Looks like all of us are still chained to Newton's clock!

Absolutes fell by the wayside when it was found that electromagnetism obeyed one set of equations and that the laws of motion, time and velocity, and energy and mass were dependent on their relative frameworks. Einstein recognized that these unknowns were important principles of knowledge about nature, and he formulated them as the special theory of relativity, which shows that absolute length, mass, energy, space, time, and motion do not exist. It describes the limits of our knowledge about reality in such a way that these limits become real descriptions of the nature of matter in relation to us, thus deepening man's understanding of the universe.

Even spookier (Einstein called it "spooky action at a distance") were the results of test after test that proved that particles communicate simultaneously at small or great distances, thus nullifying time and space as we know it. Unable to believe this, and in an attempt to disprove the results of these tests, Einstein conceived a thought experiment, known as the "EPR experiment," with colleagues Nathan Rosen and Boris Podolsky. Two quantum particles fly apart from the same origin point. Observations are performed simultaneously on both particles when they are separated (could be a yard or a million miles). The particles acted as mirror images, as in other experiments. Einstein yearned to believe that such particles have well-defined properties such as position, routing, and polarization before anyone "looks" at them. No luck!

It was not to be. The particles can behave the way they do only if they somehow are communicating across space and time, which violates all of Einstein's commonsense discoveries as well as his ideas concerning the speed of light. Einstein still rejected the idea of faster-than-the-speed-of-light signals across space and time. I am afraid that if Einstein were alive today, he would find recent experiments even spookier. Books such as *Wrinkles in Time* by astrophysicist George Smoot and science writer Keay Davidson and *About Time* and *God and the New Physics* by natural philosopher Paul Davies discuss recent theories that are even more strange.

Davies's book about God brings up issues that humankind has been grappling with for centuries: Are science and religion enemies or allies? Who or what created space-time? When and how? Was it the big bang? Was it God? Was it music? Are we asking the wrong questions? Wheeler came up with a principle he called "genesis by observership." Anything goes in the crazy world of quantum physics!

The more puzzles science solves the more they seem to proliferate. Even the principles of the big bang and of an ever expanding universe, both long held as near gospel by the majority of physicists, have come into question by such respected figures as physicist John Maddox and science writer Timothy Ferris. In *Seeing in the Dark*, Ferris probes the enigma of "dark matter," the matter that permeates more than 90 percent of the universe, countering the powerful force of gravity. Called "antigravity," for want of a better term, it somehow propels the headlong and accelerating expansion of the universe. Since gravity is the most influential of the four forces by virtue of its overwhelming prevalence in the universe, this elusive antigravity force, whatever it is, must be even more powerful. Spooky indeed!

We musicians, artists, science fiction writers, and other right-brainers need not apologize for letting our imaginations run rampant.

34

Are We All on the
Newtonian Path?

M any of us remain locked into a Newtonian outlook. In Newton's scheme of things, a body was a
self-contained unit created with an initial mass and
packet of energy that enabled it to lead a wholly independent life. Nothing would happen to it until that particle or macro body collided physically with another
body or particle. After such a happening, the two particles would continue merrily on their paths at different
angles and speeds. Such happenings are completely
subject to chance in Newton's scheme.

The Newtonian path in music ignores, downplays,
or is unrelated to historical and other contextual influences. Thus, musical form, materials, and content are
looked on as things in themselves. The path takes us
back to the views of Stravinsky, who, when I asked
about a certain passage in his music, briskly replied,

"No emotion, please, just the notes." Theodor Adorno, philosopher/musician/musicologist/Marxist and one of the most insightful thinkers of our time, took the opposite point of view, revealing historical and contemporary context and relatedness of all music.

Context: Nature versus Nurture

Context is intrinsic to the field of music. Separating any thing from its context is akin to the practice of the innkeeper Procrustes, who lopped off his lodgers' legs if they were too long for his beds! Would Mozart or Beethoven, both of whom had fathers who were musicians, have become quite so great if their fathers were innkeepers or butchers? How much did they inherit from their parents? How much did they get through their parents' genes? And how much was due to the fact that they were exposed to music from birth and the people who counted most to them were musicians? Would Einstein have achieved as much if his father and mother had not been intellectually inclined?

Scientists Richard Lewontin and Stephen Jay Gould have forcibly argued against any attempt to connect human emotions and aptitudes to Darwinian explanations. The battle was joined in 1975 when Edward O. Wilson published *Social Biology,* arguing that human behavior, like that of all creatures, was partly shaped by natural selection. So furious was this battle that a pitcher of ice water was tossed on Wilson's face at a scholarly conference in the 1970s. The authors, including Lewontin, of *Not in Our Genes* make this claim: "Sociobiology is a reductionist, biological determinist explanation of human existence. Its adherents claim . . . that the details of present and past social arrangements are the inevitable manifestations of the specific action of genes."

Steven Pinker, author of the treatise *The Blank Slate,* argues that "our brains are not general learning machines shaped entirely by culture; instead, natural selection has endowed us with a set of 'mental modules' that give us innate skills and predispositions." An example is language acquisition. The many variations of nature-versus-nurture arguments boil down to how much of our intelligence and personal characteristics are inherited or heritable, and how much they are shaped by one's life experiences.

Steven Johnson, author most recently of *Emergence: The Connective Lives of Ants, Brains, Cities, and Software,* tells us: "The most important aspect in the science of human nature is its unifying tendency, its insistence on a kind of mental toolbox shared by the most 'enlightened' westerners and the most 'primitive' hunter-gatherers of New Guinea."

Anthropologist Donald Brown, inspired by Noam Chomsky's idea of "universal grammar," documented the basic social patterns, beliefs, and categories shared by all human societies. Pinker devotes an entire appendix to Brown's list, which includes "cooking; cooperation; cooperative labor; copulation normally conducted in privacy; status; coyness display; crying; culture variability; culture; culture/nature distinction; customary greetings; daily routines; dance; death rituals," et cetera. One category not mentioned is the male-female difference, surely the subject of another book.

By Nature or Nurture?

Many people think that "perfect pitch" (the ability to identify any note that one hears) is connected with ability or even genius in music. Although there may be an element in the genes of some, there is also the ability, even in the smallest baby, to "learn" where the notes are from hearing enough music in one's family. Not born with perfect pitch, I attained it by playing an instrument and hearing lots of music. Those born with this ability have ingrained the *relative* sounds of the notes. In my experience with thousands of musicians, however, I believe that such a gift is no guarantee of musical ability or, more important, of what might be called exact relative pitch, the ability to sense precise tuning. People that have absolute pitch can often sing or play out of tune, although they know which notes are produced.

The number of vibrations or frequencies in a given period is what constitutes "pitch" in music; the larger the number, the higher the pitch. Throughout the ages, pitch standards, the criterion for tuning voices or instruments, has varied as much as one full step. The ancient Greeks tuned the kithara from a fixed pitch, approximating our A (440 hertz [Hz], or cycles per second). In 1500, the A was as high as 505 Hz, until the Church standardized it at the beginning of the seventeenth century in Germany, when it gradually lowered considerably, only to rise again at the beginning of the eighteenth century. It was just under 430 Hz during the lifetimes of Bach, Haydn, Rossini, Purcell, Mozart, Beethoven, and Handel. The French raised it to 435, and the Americans went as high as 445. 440 is the standard today. The "nature" of music, like everything else, is subject to change.

Dr. Diana Deutsch has studied the tones that are found in speech and has found that Vietnamese and Mandarin speakers use identical tones for the same words, even if measured days apart. This is a kind of perfect pitch as is understood in music. "It is still possible that the subjects may not see or realize a connection between tone as they use it in language, and pitch as a musical concept," said Dr. Donald Hall, a physicist at the California State University in Sacramento, a church organist with perfect pitch.

Other research has shown that the prevalence of perfect pitch is higher in Japan, where, although the language is not tonal, many children receive Suzuki music training. Perfect pitch is also more common among professional musicians, but this may show merely a predisposition to the trade. Recently, some scientists have contended that most babies are born with perfect pitch but retain it only by learning a tonal language such as Chinese or by undergoing early musical training.

"There can be a much higher incidence of absolute pitch musicians out there if all of us were exposed to music much earlier," said Dr. Gottfried Schlaug, a neurologist at Beth Israel, Boston, who has studied the relationship of structure in the brain and perfect pitch. Perfect pitch can be nurtured and self-trained. It would benefit children to have early music exposure, both to develop neuronal pathways for the comprehension and appreciation of music and to augment other skill sets, such as math.

35

Creativity

*"Science . . . means unresting endeavor and
continually progressing development towards an
aim which the poetic intuition may apprehend, but
which intellect can never fully grasp."*
— MAX PLANCK, father of quantum mechanics

The opposite of destruction is "construction," often aligned with "creative," a word overused, misused, and maligned. "Creativity" often has the odor of Hollywood or Madison Avenue surrounding it. One may rightfully be suspicious of the word. We do not create out of thin air. I have even been tempted to exaggerate and say there is no such thing as creativity. Instead, the creative person selects what already exists and then arranges the notes or sculpts the marble "cooperatively" with the surrounding materials on earth or with the forces of the universe all about us. The "geniuses" are those who select well and then rearrange in order to satisfy their vision, which, in the case of Beethoven or Michelangelo, becomes our vision and our art.

From the very beginning of science, the cause-and-effect relationship has been recognized. Everything emerges in the confluence of time from those people, discoveries, and things that led to the next event. As we step on the threshold of modifying and actually "creating" life, the temptation to crow about human creativity is irresistible. Cloning and technical advances in health care and surgery have tempted us to play God. Humility has taken a backseat in science, philosophy, and politics. Trying to find a scintilla of humility in politicians is like trying to find my wife's earrings at the bottom of the ocean.

Pace

As in music, nature has a certain pace that can be too suddenly hastened (or retarded) at our peril. Scientists think that we are moving too fast for nature to catch up and that we are destroying hundreds of species and the earth's ecology in the process. How radically can we be creative— whether biologically or otherwise? Change occurs organically from preceding elements and developments, which themselves must have obeyed a natural, causal pace. This is not to deny the uneven pace of societal, scientific, or individual development and change.

Some children learn more quickly than others, becoming Itzhak Perlmans or Albert Einsteins. But scientific discoveries usually arrive after long, slow, and tedious growth, until a sudden flash of insight from previous data brings science to an utterly new level. Society too proceeds by jumps and starts, changing conditions gradually heating up until a revolution, small or large, brings sudden, radical change.

We know from physics that everything emerges from everything else. Creativity, as commonly used, comes close to denying change. The big-bang theory of the creation of the universe is one such denial, and may be the reason that some religious fundamentalists, who need an escape from much-battered creationism, have happily co-opted the big-bang concept as a quasi-scientific justification for their religion.

A recent book, *Intelligent Design Creationism and Its Critics,* edited by Robert T. Pennock, has elicited much response, both pro and con. "Intelligent design" means design according to what may be called cosmic blueprints, but blueprints that humans are fully able to read and decipher. Humans thus become godlike, creative architects.

The Jury Is Out

So what about the "many big bangs" and "many universes" theories? Though the evidence is weighted on the side of the one big bang, there are a sufficient number of authoritative dissenters to indicate that the jury is still out, though leading scientists agree on continual process and change in the universe. Nobel Prize winner Isador Rabi wrote, in 1975, "I don't think that physics will ever have an end. I think that the novelty of nature is such that its variety will be infinite—not just in changing forms, but in the profundity of insight and the newness of ideas."

Creativity and the Arts

Dr. F. X. Barron conducted intensive studies of highly creative people at the Institute for Personality Assessment and Research at the University of California in which creative people, ranked for their originality by their peers, spent four or five days at the Institute for extensive psychological testing and interviews. He concluded that highly creative people are "more primitive and more cultivated, more destructive, a lot madder and a lot saner than the average. . . . They resisted authority and were willing to take risks. They had a high tolerance for disorder and a preference for complexity, combined with the ability to extract order from chaos. They also showed high levels of ego-strength." We might add the prime roles played by family background, historical period, culture, and context.

Creators or Arrangers?

Mozart, Beethoven, or any other composer of the period could not have written the same music several centuries before their births. Obviously, the musical materials they employed had evolved to the point where they could arrange these materials as they did. Beethoven was not merely a creative arranger—his arrangements were *extraordinary*. He took the same materials—such as notes, scales, modes, and intervals—as lesser lights, and put them together in such a fashion that they produced high magic and unequaled symmetry.

The reader would be well rewarded to take a look at the fascinating, easily obtained Beethoven notebooks in which he jotted down ideas for his compositions. In his Fifth Symphony, for example, he took a two-note theme, and, like a neutrino that changes form or a particle that becomes an antiparticle, he created what is essentially a theme and variations. In the process of that creative arrangement, Beethoven tried many approaches, working and reworking the notes, themes, and phrases until he finally got it right! We are the lucky recipients of the finished product, which seems as though it could not be any other way. He wrote four different overtures to one opera, *Leonore,* all of which are published, though one is the most popular with conductors and audiences. (Listen to them and choose for yourself!) What an arranger was he! A legitimate question, of course, is why the many composers (or painters or choreographers) who share a particular era, all of whom have at their disposal the same notes, keys, modulations, paint colors, etc., are not all Beethovens, Bernsteins, Botticellis, or Beatles!

The answer is that all art, science, philosophy, politics, social structure, and culture are adaptations to the music of the universe. This is true of any creative act or idea whose time has come. Listen to Benjamin Britten's *War Requiem* or *Serenade,* Charles Ives's Fourth Symphony, or Olivier Messiaen's *Music for the End of Time.* Luciano Berio reveals examples of contemporary reconstruction and deconstruction of previous works in *Coro* and *Una vera Storia.* Classical composers of the eighteenth and nineteenth centuries unashamedly incorporated folk tunes or themes from fellow composers with no thought of plagiarism. Many plagiarized themselves, recycling their favorite themes in work after work. Ives took creative plagiarism to a new level; he used everything from band marches (such as "Columbia, Gem of the Ocean") to common tunes (such as "Hello My Ragtime Girl" and "America") and arranged them in such a way that they became wholly new. Mozart and other composers wrote arrangements of Handel's famous *Messiah,* adding instruments and making other changes. Handel availed himself of whatever instruments and players he could get for a particular occasion. This puts to death the current arguments among lofty musical historians rigidly espousing this or that orchestration of the *Messiah* for today's performances.

In physics as well as in other disciplines, genius arrives to stand on the shoulders of those who have come before. This allows for a better view of the horizon, where what can then be produced is a historic insight that echoes forever as part of the music of time.

36

Context and Frames
of Reference

M usic is a social art, whether defined in the traditional sense or defined broadly as in these pages.
When I first became involved with the United Nations
back in the 1970s, I invited a group of African delegates
to my home and, during the course of the evening,
asked if they would demonstrate their music. Expecting
songs or even chants, they instead arose, formed a circle, and danced. Dance to them was music. Only later
did they give vocal expression to their music!

Music does not exist apart from the interrelatedness
that shapes our lives. It cannot divorce itself from the
people who wrote it or those who hear it. Shakespeare
had it right: It does indeed "hold the mirror up to nature," nature in the human sense and nature in the cosmic world of science. It imbues the plant and animal
world and penetrates even the inorganic world from the

smallest particle (a superstring?) to the largest cluster of galaxies and beyond. Twentieth-century physics has repeatedly confirmed by experiment that which we know by intuition. Music reflects the longings, aspirations, humor, tragedy, and spirituality that give meaning to our lives; it need not enter these lofty realms to be a social art. It arises and flourishes in the social and physical medium of its time and culture.

Any fragmentation of music from this context can exist only as a temporary phenomenon. Philip Glass's music, in constant motion—contrapuntally and harmonically—provides an excellent example of the birth, growth, and completion (death) of a musical work. If you try to fragment any part from the whole, the result will be frustration. By the same token, any attempt to freeze the process of change, whether in a work of art or a societal system, has a very limited existence before a shift inevitably takes place, minor or tectonic.

The concept of context is opposed to fragmentation. Context joins together. It sees a part in relation to other parts and to the whole. The central content of religion or the mystical experience consists in a revelation of cosmic unity. Plotinus defined illumination as "absolute knowledge founded on the identity of the mind knowing with the object known." In mystic states we become one with the Absolute, and we become aware of our oneness. Any part of a song or symphony must fit precisely within the context of every other part and of the whole. Each object contains the seed of everything else.

> An Indian sage wrote, "In a grain of dust are all the scrolls of the sutras in the universe: in a grain of dust are all the infinite Buddhas. Body and mind are together with a blade of grass and a tree. Therefore, the One Mind is all things; all things are the One Mind."

William Blake put that concept into the first stanza of one of his poems:

To see a World in a Grain of Sand
And Heaven in a Wild Flower

Hold Infinity in the palm of your hand
And Eternity in an hour.

Space is another concept that has been put into context since Einstein's general theory of relativity. Ancients thought the earth was flat because, from their frame of reference, it certainly looked flat. But when it was discovered that the earth was round, we realized the limitation of our perspective. Our perspective is still badly limited, though in different ways. It feels like our lives are out of sync with the universe and that the universe is talking back. Having been the evacuee victim of three hurricanes in a few weeks, I can attest to Nature's revenge.

In observing curved motion, we do not distinguish between acceleration and gravity. To make life simpler, then, we dispense with gravity altogether. If we observe any curved motion, we can blame it on the acceleration of our RF (reference frame) and need not invoke any mysterious, invisible gravitational field. Even the distinction between IRFs (inertial reference frames) and RFs turns out to be a red herring. Inertial motion appears straight in an IRF and curved in a non-IRF. These appearances are not inherent in the motion but are associated with the RFs themselves. Motion appears curved in an RF that has curved geometry; straight in an RF that has flat geometry. IRFs constitute a special type of RFs that have flat geometry, which is the case in special relativity. General relativity treats all cases—of motion and physics in any kind of RF, which in general has curved geometry.

Good old Newton concocted a force of gravity to explain departures from the laws of inertia. Newton reasoned that Mars, if left to its own devices, would follow a straight path at constant speed. An isolated body slavishly obeys the laws of inertia and moves uniformly. Because we never do see Mars obey the laws of inertia, Newton thought that something must cause Mars to move on a curved path. He called that something the force of gravity. But all that Newton or anyone else ever observed was the motion of Mars—never the force of gravity.

Einstein stated that "the laws of physics are much simpler than Newton imagined. We needn't explain the departures from the law of inertia. There are none! All motion is inertial. Objects simply follow the natural contours of space-time. The shape of those contours depends on the RF in which they are viewed. In an IRF, the contours happen to be straight lines. In other RFs, the contours are curved. There is no essential or physical difference between IRFs and other RFs, only a geometrical one. The laws of physics are the same in all RFs, and all motion is inertial."

In a famous experiment, Einstein imagined himself in an elevator. He wondered why he was standing on the floor. It seemed obvious and naive, and yet the answer is found in one of the best kept secrets in science. One would say that it is the gravitational pull of the earth that held him there. But Einstein was not satisfied:

> All well and good, but I can't see out of the elevator. How do I know I'm really on the earth? After all, there's another way to explain why I'm standing on the floor. My elevator could be in outer space, hitched to the back of a rocket ship that is accelerating upward. As the elevator accelerates upward, I am pressed downward toward the floor, just as happens on the elevator on the earth. And as I accelerate upward, if I release a key from my hand, it too will drop.
>
> As soon as I let go of the key it stops accelerating, but it continues to move at the speed it had when I released it. This is just a law of inertia in operation: bodies always move at constant speed unless that force accelerates them. As in empty space, far from any planets or stars, there is no significant force of gravity to accelerate the key once I let go of it.
>
> Suppose the elevator is moving 100 miles an hour when I release the key. A second later, I move at 101 miles an hour because I'm accelerating, and the key continues to move at 100 miles an hour. The elevator, with me in it, continues to accelerate to 102, 103, and so on, while the key once released no longer accelerates, it keeps moving at the same 100 mph. The key then "falls behind" because the elevator floor "catches up" with the key. Eventually the floor

reaches the key, but to me it simply appears that the key "falls" to the floor.

Whether I am on an elevator at rest on the earth's surface or in an accelerated elevator in space, I see and experience the same thing. In each case I am held firmly against the floor, and the key falls at the same rate. Unless I can see outside my elevator, I cannot tell whether I am in the earth's gravitational field or accelerating in space. Now, of course, if the rocket starts, stops, or accelerates or decelerates, then I as the object in the elevator would move differently and could easily tell I am not on the earth. But if the rocket accelerates smoothly and at a constant rate, then no experiment I perform or observation I can make, can distinguish between the two cases. From an empirical point-of-view, they are equivalent.

This is how Einstein discovered, by way of this and other experiments, the remarkable equivalents between gravity and acceleration, which he described later as "the happiest part of my life." This connection became the cornerstone of his general theory of relativity, which he called the principle of equivalence. The "frame of reference" is the seminal idea of his general theory of relativity.

We have seen that gravitational fields or space-time affects clocks. Einstein also predicted that a ray of light would follow a curved path in a gravitational field. For example, he calculated that in passing near the sun's surface the light from the star on its way to the earth will be bent or deflected a tiny amount. Sir Arthur Stanley Eddington, a famous English astronomer, confirmed this. The British government funded an expedition to observe the sun's eclipse in 1919. Eddington and his colleagues observed the eclipse from opposite sides of the Atlantic Ocean resulting in confirmation of Einstein's prediction. This gave Einstein instant fame. In the December 1919 edition of *Berliner Rustingerzeitung,* a cover story and picture appeared with the caption "A great new figure in world history, Albert Einstein, whose investigation signified a revision of our concepts of Nature and our empowerment with the insights of a Copernicus a Kepler and a Newton." *The Times* of London had similar proclamations. However, it took decades before the full significance of his discoveries could be applied and explained. The experiments he performed to test the many predictions of general relativity enabled us also to test accurately the radio waves from quasars and from very remote objects in the skies. This eased the space mission to Mars.

37

Is the Universe Really
Made of Music?

M aybe the future will bring rewards to theorists in the other field of untold mysteries, music! As we have seen, instrumental strings have played a key role in wave theories and mathematical formulations for thousands of years, and according to some scientific explorers, music predates civilization and the emergence of life itself. We are beginning to learn that so-called inorganic matter is not so "lifeless" as we thought. If superstring theorists reveal that all matter in the universe is based on these strings, then why not postulate that the universe is one big orchestral string section?

In quantum mechanics, energy is frequency multiplied by Planck's constant. Planck's constant in this context might be roughly analogous to the frequency of the lowest note (G) on the violin. Of course the scales are enormously different. Superstrings could be visualized as being in the lowest mode while heavier particles occupy the higher modes on to infinity.

The "super" in superstring theory indicates "supersymmetry," a subject I shall handle warily, for this is an esoteric subject that has eluded me. Even some of the most advanced physicists find themselves scratching their heads on this one. The conclusions drawn from supersymmetry are understandably controversial among scientists and laypeople alike.

Columbia University physicist Brian Greene is, in my view, like the late Leonard Bernstein, whom readers may recall as one of the great conductors, composers, and teachers of the twentieth century. His televised explanations of the complexities of music painlessly and lucidly educated generations of listener/viewers, who then became aficionados of classical music. His books, *The Unanswered Question* and *The Joy of Music* among them, are a revelation. Brian Greene's bestsellers, *The Elegant Universe* and *The Fabric of the Cosmos,* explain string theory and physics as well as Bernstein did music. From my prejudiced point of view, Greene had a much more difficult subject. I highly recommend his books.

Musical Symmetry

The concept of symmetry is not confined to string theory or physics or math. The concept is everywhere about us. It holds an honored place in history and is related to many disciplines. Plato's *Theaetetus* examines the eurhythmic planning system, while the classical meaning of the word "symmetry" includes the Pythagorean creed that "everything is arranged according to number." This is not only justified in art, but it has also been used in biology and, of course, in geometry and in other forms of mathematics.

Symmetry is found in spirals, in all living forms, and in the proportions of architecture, from Greek and Gothic structures to modern ones. The granddaddy of symmetry research is Kepler's *The Divine Proportion.*

In music, symmetry is easily demonstrated by way of tempo, rhythm, melody, and harmony. This is in all of music, even in those ultramodern works whose composers might deny it.

The compositions that remain popular through the years are those that clearly elucidate proportion and symmetry. Beethoven's Fifth and Ninth symphonies are both based on the simplest of materials, a descending interval of a third in the case of the Fifth Symphony and open fifths and octaves that open the Ninth. Much of the remainder of both works is splendidly symmetric as to rhythm, melody, harmony, and placement. If examined carefully, one can analyze virtually all the variations and permutations on the basis of symmetries or asymmetries. The works of Mozart, Haydn, and most classical compositions are highly symmetrical. Even those seemingly complex works by Stravinsky, such as *The Rite of Spring, Petrouchka,* and *The Firebird,* can be boiled down to basic materials of music. Even when buried in ingenious formulations, they can still be dug up. Brahms's Fourth Symphony turns the theme of a descending third into its mirror image of an ascending sixth with these intervals appearing vertically (in the chords and harmonies) as well as horizontally (the melody). Historically, a simple modulation, a temporary discord, or a delayed resolution could have been experienced as a pleasant shock by the ears of the time. These days, it takes a great deal to shock ears accustomed to the omnipresent roars of boom boxes, traffic, jackhammers, and other high-decibel abominations.

Gustav Holst (1874–1934) bridged the gap between the "true" Romantics and the Modernists. In *The Planets* (1915), he employed a huge orchestra with organ and chorus, and utilized an arsenal of techniques that came eventually to be adopted by others. The work's first movement, "Mars, the Bringer of War," was sketched in the weeks and months preceding World War I. A Romantic at heart, Holst was responding to the increasingly hostile climate of the time, and I conjecture that he was also interested in the implications of the "harmony of the spheres." His use of the tritone (the augmented fourth interval, forbidden in counterpoint and historically condemned by the Church) in the movement's conclusion indicates a fundamental break with previous practice.

Twentieth-century composers, as exemplified by Arnold Schoenberg, Alban Berg, and Anton Webern, claimed to liberate music from the dictatorship of the harmonic series, while imposing stringent compositional requirements. Others also tried to break the bonds of European music, the harmonic series, and symmetry. A twenty-four-year-old George Antheil wrote *Ballet Mécanique,* featuring three airplane propellers and other odds and ends. For a while he became the darling of fashionable patrons and the bourgeois intelligentsia as well as those writers and artists who inhabited the 1920s and early 1930s. When I met him in the late 1940s and early 1950s, he offered to compose a work for me, but I showed little enthusiasm and the deal was never consummated.

George became the Eminem of his time when he announced: "Now I hope to present you with . . . the physical realization of the fourth dimension. I am not presenting you with an abstraction. I am presenting you with a physicality like sexual intercourse." Henry Cowell, a more serious composer/teacher and a pioneer of modern music, became famous for the clangor of his piano writing. In that period, asymmetry ruled or misruled, depending on your point of view.

Dr. Val Fitch, the physicist who won a Nobel Prize for his work on the tiny asymmetry in the weak force that made it possible for our universe to exist, was head of the physics department at Princeton University. He published, together with his colleague James Cronin, the astonishing discovery that the weak force violated the balance of positive and negative forces in the pre-universe, allowing the creation of the universe. A story about Fitch has it that when housed in a dreary Russian hotel during a conference, which included such famed luminaries as Murray Gell-Mann, Fitch introduced some of his hosts to an American technology, the Frisbee!

Proportions in space and time have been glorified among mathematicians and philosophers throughout the centuries. We look with awe at the symmetrical proportions of the Egyptian pyramids, whose builders must surely have studied symmetry. Leonardo da Vinci's designs and plans are based on grids of symmetry. Mystics such as neo-Platonists and Pythagoras's heirs have taken heart from the Pythagorean doctrines of numbers and symmetry. For them, the "divine" numbers are the archetypes, eternal patterns that guide the Great Orderly One. The different integers were not simply 2, 3, 4, 5, etc., but the dyad, triad, tetrad, pentad, decad (10), etc.

The mystics in the Hebrew Kabbalah and the Alexandrian diaspora had other associations for numbers. Their system consisted of an extremely complex web, which could take a full book to describe. For example, 5 is the number of love, and 7 of the virgin. To approach the holy site of Jerusalem in ignorance is to encounter danger. But if one approaches in such a way as to harness its metaphysical power, the possibilities are limitless. Once there, the visitor must begin by reciting the thirty-third Psalm in Hebrew, correctly pronounced. Numerical values are attached to the letters of the text. The Psalm consists of 22 verses just as the Aleph Beth (ABC's) consists of 22 letters, and those 22 verses are comprised of 161 words. The number 161 corresponds to the biblical code of Eyeh (104): Aleph (111), Hey + Yud = 15, Yud + Vav + Dalet = 20, and by some kind of complicated math that sounds like legerdemain to me, the original code adds up to 161—the number of words in Psalm 33. But the Kabbalah is hardly alone in assigning numbers symbolically, or, in the case of science and computers, realistically!

Music is as pervasive in the other arts as it is elsewhere in the universe. It is not uncommon to hear architecture spoken of as "frozen music." Rarely is dance seen without music. Somewhat less common is the description of painting in terms of color vibration based on the wave frequencies emitted by each color. I once worked with an inventor who developed a device that translated musical sounds into colors and shapes. We set up a screen behind my orchestra, and he projected the images that reflected the music's dynamics and "colors." One could also describe painting and sculpture as music. When I was in my early twenties, I visited the smaller St. Peter's Church in Rome to see Michelangelo's *Moses.* I stood before this huge sculpture in the dim light of the church's basement and literally shook in reaction to its magnificence and power. What music Michelangelo made!

Perhaps poetry, more than architecture, painting, or sculpture can be most easily seen as music, for it is usually "right brain" and abstract as is music. Poetry's rhythm, alliteration, symmetry, and other forms or non-forms are highly congruent with music and musical form. Some well-written prose is sometimes alluded to as "music to the ears" ("eyes" might be more appropriate, though I do hear the music in my head when I read a score or a book by an author who has special appeal).

Word Music

O Rose, thou art sick!
The invisible worm
That flies in the night,
In the howling storm,
Has found out thy bed
Of crimson joy:
And his dark secret love, Does thy life destroy.
—WILLIAM BLAKE

William Blake, artist, poet, engraver, and political radical in eighteenth- and nineteenth-century England, is one of the few poets I look on with the same reverence that I accord to great composers, and for similar reasons. To me he is a musician, though, as far as I know, no literary historian has ever attached this description to Blake. Another poet/playwright whose "music" moves the world is William Shakespeare.

Blake's overwhelming feelings turned his visions into poems, engravings, and paintings, all interconnected. His *Songs of Experience* are poems about fear, deceit, and cruelty. Accompanying engravings look like poisonous roots and coils encircling and confining the script. Words and image combine to make a complete language, much like polyphony in music. *Songs of Innocence* features hymnal and lullaby rhythms, with lessons addressed to "every child." The art includes cherubic boys and girls, pure and naked. The little chimney sweep, in heaven, released from his coffin by an angel is depicted as a dark mass on a dark London street. His "Tyger! Tyger! burning bright, / In the forests of the night" is a somewhat whimsical catlike figure, hardly fearsome despite the lines, "What immortal hand or eye / Could frame thy fearful symmetry?"

Under the heading "Fearful Symmetry," Margo Jefferson, in a deeply revealing *New York Times Book Review* piece ("Writers and Writing"), comments on the Blake exhibition at the Metropolitan Museum of Art in New York. Her reference to symmetry is that which exists between art and poetry. I don't think she would object if music and even physics were thrown in.

Ms. Jefferson composed a bit of music herself in her closing lines when she describes the geometry in the "perfectly aligned writing, . . . ironically scientific illustrations" in Lewis Carroll's 1862 manuscript, *Alice's Adventures Under Ground.* "Alice is a dreamier figure here. Her hair is looser. . . . The animals are more eccentric and touching . . . [they are] highstrung. . . . The animals she swims through the pool of tears with, the hookah smoking caterpillar, look more fantastic but more intimate too. Alice is not making her way boldly but primly through Wonderland; she is finding her way through the underground of the imagination."

In his *The Unifying Power of Art,* Jean-Pierre Changeux asks, "What goes on in the mind of an artist when he is creating? What is the origin of aesthetic pleasure and can it be explained?" Director of the Pasteur Institute and professor at Le Collége de France, M. Changeux is not only a scientist but also an art lover and collector of paintings. As president of the Dating Commission of the National Center of Fine Art, he is in a good position to express his reservations about contemporary art and artists.

38

Molecular Friends and Enemies

A scientist, who worked on the very small rather than the cosmological, discovered some societies smaller than human scale. In an article entitled "Circling the Wagons: Molecular vs. Human Behavior," Seymour Meyerson, a scientific and social generalist as well as a student of sociology, social psychology, and comparative religion for more than fifty years, describes a series of experiments on the sociological behavior of molecules. According to Meyerson, much molecular behavior seems to parallel the sociology of humans and animals. For example, when an individual molecule is separated from its neighbors, it may, upon excitation, "bite its own tail or otherwise reorganize itself and/or go to pieces."

An oil molecule will, upon encountering another of its kind, immediately interact with it. "For example, the lubricating oil molecules in your automobile are markedly affected by temperature because of the associated change in viscosity. To compensate for such a change, oil formulators add VI (viscosity-index) improvers. These additives consist of very large flexible molecules with chemical characteristics differing markedly from those of the oil molecules, to which they minimize their exposure by coiling into tight balls, effectively taking themselves out of solution. If two such molecules happen to meet, they will uncoil and coil about each other, or a sufficient rise in temperature will induce them to uncoil even without a partner. Such uncoiling effectively puts the VI improver molecules back into solution, thereby increasing the viscosity of the oil and thus compensating for the viscosity-lowering effect of the rise in temperature."

Molecules react like a crowd at a soccer game. Many individuals in that crowd may act in ways that they never would outside of a soccer stadium. But at some signal, let us say a controversial call or a fight among players, the crowd might become a rowdy mob. Police action may actually increase the violence, or a plea from the loudspeaker might "uncoil" the unruly fans.

A molecule, isolated in space, will seek out molecules with similar chemical characteristics. The molecules seem to distinguish "friends" like themselves from those with different properties. Think of the Protestant-Catholic conflict in Ireland, or other all-too familiar examples.

Musical Electrons and Molecules

When a little energy is added to any metal, the electrons move faster to the point of chaos. At a critical point the chaos is transformed into collective behavior. Like musicians, electrons exhibit both individual and coordinated behavior. In some metals, when a critical low temperature (close to absolute zero) is reached, those metals are suddenly transformed into superconductors, where all resistance ceases and current has the ability to flow for thousands of years. This phenomenon is based on a form of collective order. This radically new form of behavior arises from the cooperative action of the whole. Every musician knows that the temperature in the hall affects his or her playing, as does the size and character of the audience. The instruments are similarly affected by temperature and other environmental factors.

A lonely molecule will, with great energy, seek out its community of molecules as soon as possible. In the presence of great numbers of other molecules, the molecule exhibits behavior not unlike that which superstring-theory physicists describe, that is, coiling and uncoiling until all parts of the molecule are in close contact with the same or neighboring molecules. A collection of similar molecules, looking out at an environment that it perceives to be different from itself and hence unfriendly, tries to attain a configuration that will minimize the surface exposed to the outside world.

Meyerson hastens to point out that "in human communities, unlike molecular ones, we have the capacity to transform a 'poor solvent' [environment] into a 'good solvent' by purposeful action and we do not need a rise in temperature to prompt us to 'uncoil.' Ultimately we can expand the group in which we feel at home to embrace the entire human race. *We* have choices."

Consider the random warming up of an orchestra. Each instrumentalist plays a few notes or a bit of musical passage different from all the others. Each string player may pick or bow the strings, some up-bow, some down-bow. The woodwinds, brass, and percussion add to the melee until the warming-up stops by a signal from the concertmaster, who stands and then indicates for the oboe to play an A, and each section tunes, in turn. Then, the conductor arrives. His eyes and baton, in relatively small signals, usher in collective motion, rhythm, and harmonically unified sound. Like an aberrant molecule, a mistake by a musician or an infrequent mistake in the score can disrupt the behavior of the entire orchestra.

The symphony orchestra may be the best paradigm for society as a whole. But that subject belongs to another chapter, as does the extrapolation of molecular behavior to political elections, strikes, and revolutions. In the world of humans, where we fear or hate "the other," the "us versus them" syndrome is all too clear.

I would surmise that Meyerson, like Einstein, would weigh in on the side of causality, as opposed to the ideas of some contemporary philosophers who would probably argue that everything is random and chaotic, allowing little or no choice as we drift with the winds of chance.

Molecular Psychology

In an award acceptance speech, "Reflections on the Psychology of Molecules," Meyerson pointed out, although clearly with tongue in cheek, that "all of us who work with mass spectrometers know that these machines are human, at least in some respects. I want to argue today that the same can be said of the molecules that we use the machines to observe."

This view enables one to look at both molecules and people from unusual angles, which may contribute some added insights to both contexts. Anthropomorphism traditionally has not been fashionable in the scientific community, but perhaps "the times they are a-changin'."

I would like to think that blinders are gradually being lifted from the eyes of members of the science world. Indeed, this growing awareness among scientists is being shared by the general population. The exigencies of modern life serve as an excellent teacher. Medical doctors, for example, have been forced into a more holistic approach as their patients insist on the use of alternative modalities of disease prevention and cure and as drugs are increasingly called into question.

39

Alchemy: The "Black Magic"

A long time ago, what we call chemists were called alchemists. Their working assumption was that everything on earth should be made up of a single element. The alchemists worked feverishly to transmute base metals into gold, but their motives were not only pecuniary. They believed that if you knew about the fundamental substance, you could also develop a cure-all for disease. They spent their entire lives in these pursuits, writing reams of notes in Arabic and Greek. These labors were not unlike those of scientists today, as physicists race to unify the four basic forces in the universe and find a theory of everything. The alchemists did accomplish some important things. As more and more workers became involved in thousands of experiments over the centuries, hard facts emerged about the behavior of metals and alloys, acids, bases and salts, and the mathematics of thermodynamics. It wasn't all black magic.

Paracelsus, a sixteenth-century physician, became famous for his medical use of mercury and arsenic. Three centuries of patients were treated with potions of metals for all kinds of illnesses. It is interesting to note that arsenic and colloidal silver are being revived today to treat certain varieties of cancer and other ills.

Who Is What?

In the twentieth century, astronomers had to become physicists and physicists became astronomers. Both became chemists, since their work often touches on chemistry. When I studied chemistry in high school, the periodic table of all the elements (at that time we knew of about ninety) was largely the purview of chemists, but it was already leaking over to the field of physics. Now a study of the elements is elementary for all scientific disciplines. Today, all scientists study the interrelationships of the various elements with each other and the qualities of the resultant compounds.

Wild Dreams and Nightmares

The alchemists didn't have the benefit of the stunning discoveries of the recent century or two. Our contemporary "alchemists," who inherit the entire magnificent legacy of science, are the brilliant, mostly young physicists who spend their time seriously working on theories of interstellar space travel, wormholes, many universes, the bio-sciences and backward and forward time travel.

These schemes are not bad in themselves. From time immemorial, great discoveries emerged from dreams just as wild. These talented scientists want to make their mark in the world. They know that one of the best ways to get attention and rewards in today's media-mad environment, with the attendant positions and funding, is to work on schemes such as these. These dreamers might usefully point their research toward, for example, tackling ways to avoid tornadoes or determining the effects of genetically altered foods on health (beware of connection with the hugely profitable chemical industry!). Prevention of disease without surgery or drugs could be a most fruitful investigation.

David and Goliath

Exercise and fitness trainers are doing very well while vitamins and herbs have become big business. Even the tiny organic food farms are growing, as the giant industrial farms and pharmaceutical companies take advantage of the profit possibilities. David is testing his sling in preparation for an attack on Goliath. Traditionally conservative small farmers and fishermen are being phased out as their farms are taken over by conglomerates. Fish farming and chemicals in oceans and rivers has resulted in one species of fish after another becoming endangered or extinct. And beware!: Young scientists facing an uncertain future with basic research funding drying up have turned their attention to biotechnology and biomedical research, a favorite of the pharmaceutical industries.

The Uncertainty Principle
in Music

The arts community is even more vulnerable to cost cuts than are the sciences. Many orchestras and arts groups are closing shop as they face mounting deficits. Bankruptcies are common among small and mid-level orchestras and dance groups. Musicians and arts groups are joining research scientists in being forced to find income from other sources.

In any case, change is blindly accelerating. There is no precedent in history to guide us, unless it is the lessons of World War I that led to

Hitler's Germany. Hopefully, other alternatives can be found. Can science and music play a role?

Copenhagen, a broadway play (from Tom Stoppard's *Arcadia*) about two of the foremost physicists of the twentieth century and their conflicts and discoveries, received rave reviews. It tells of a fabled meeting between Niels Bohr, the great Danish physicist, and Werner Heisenberg, the German physicist who spent the World War II years serving the Nazi regime as its leading scientist. Heisenberg had been a student and protégé of Bohr's and was treated like a son in the Bohr household. Playwright Michael Frayn explains the background: "The central event is a real one. Heisenberg did go to Copenhagen in 1941, and there was a meeting with Bohr, in the teeth of all the dangers and difficulties encountered by my characters. He almost certainly went to dinner at the Bohr's house, and the two men almost certainly went for a walk to escape from any possible microphones, though there is some dispute about even these simple matters. Worse disputes have surrounded the question about what they actually said to each other, and where there's ambiguity in the play about what happened it is because of the recollection of the participants. Much more sustained speculation still has been devoted to the question of what Heisenberg was hoping to achieve by the meeting."

Did Heisenberg hope to extract secrets from Bohr about the war plans of the U.S. and Britain? Did he simply want to renew acquaintance with his beloved teacher? Is the same perhaps true in reverse for Bohr, that he wanted to see the man who he felt was another son to replace his lost son? And what role did Bohr's wife play? Heisenberg says in his memoirs, "We both came to feel it would be better to stop disturbing the spirits of the past." How true is this statement? Heisenberg lived under a cloud of suspicion after the war, lasting until his death years later. Was he a Nazi, or did he subtly try to undermine the German war effort? In any case, he was treated with hostility and contempt.

Frayn reveals in his play that Heisenberg was a German patriot, but we still wonder what role the scientist actually played. Frayn continues, "I can't claim to be the first person to notice the parallels between Heisenberg's science and his life. . . . Many different views about this man and his behavior have been expressed, views that have been fervently, even passionately held by a variety of individuals. It is as if the intense emotions unleashed by the unspeakable horrors of that war and that regime have combined with the many ambiguities, dualities, and compromises of his life and actions, to make Heisenberg himself subject to a type of uncertainty principle."

Heisenberg and the rest of the German atomic team were interned toward the end of the war in a house bugged by British intelligence. The recorded conversations were kept secret by the British government for "reasons of state" in spite of the most strenuous efforts by historians. A combined assault by the leading members of the Royal Society in 1992 finally pried them loose. But uncertainty continued.

It is true that the concept of uncertainty is one of those scientific notions that has become generalized to the point of losing much of its original meaning. The idea as introduced by Heisenberg into quantum mechanics was precisely defended. It did not suggest that everything about the behavior of particles was unknowable or hazy. It was limited to the simultaneous measurement of connected variables such as position and momentum or energy and time. The more precisely you measure one variable, it said, the less precise the measurement of the related variable can be; and this ratio, the uncertainty relationship, is itself precisely formulable.

Einstein tried to deny the uncertainty principle, setting up a series of unsuccessful experiments called EPR. To the end of his days, he never quite accommodated himself to the uncertainty principle and even to aspects of quantum mechanics, which gave rise to it. Physicist John Wheeler explained: "We had this old idea, that there was a universe out there, and here is man the observer, safely protected from the universe by a six-inch slab of plate glass. Now we learn from the quantum world that even to observe so minuscule an object as an electron, we have to shatter the plate glass; we have to reach in there. So the old word, 'observer,' simply has to be crossed off the books and replaced with the new word, 'participator.' In this way, we come to realize that the world is a participatory universe."

In all the years since my talk on our high school lawn with my fellow fourteen-year-old, I think I was, on some barely conscious level, struggling, like Einstein, with the paradox of relativity and the "participatory universe." Now I am convinced that both concepts are right!

Heisenberg was under contradictory pressures during and after the war that made it particularly difficult for him to explain what he was trying to do. He wanted to distance himself from the Nazis, but did not want to suggest that he had been a traitor. He was reluctant to claim to his fellow Germans that he had deliberately lost them the war, but he was no less reluctant to suggest that he had failed them simply out of incompetence.

But Heisenberg was subject to pressure long before he needed to explain himself, and the uncertainty about his intentions begins much earlier and goes much deeper. Perhaps the real question is what he knew himself about what he was doing. This paradox in the play is mirrored in real life. When any of us look into ourselves, we come up against limitations of what we can observe and can know that are not entirely unlike the barriers that he and Bohr had identified that restrict our knowledge of events inside the atom, barriers that limit our understanding of ourselves and the world we inhabit. I have found the physicists that I have met, read, or interviewed to be intensely human and, like musicians, passionate about their art. All are excited by the drama inherent in physics, the probing of the deepest secrets of life and the universe. In their different ways, both music and physics try to explain the meaning of life.

Heisenberg remembered conversations with Bohr late at night in 1927 that ended "almost in despair." He repeated to himself the question, "Could nature possibly be as absurd as it seemed in these experiments?" He was talking about quantum mechanics and how it presents a world that seems upside down according to our perception.

Einstein objected to quantum mechanics because it uses probabilities instead of certainties, and he did not see this as a valid foundation for a total theory. "Quantum mechanics is very impressive," he wrote, "but I am convinced that God does not play dice." Becoming "an obstinate heretic in the eyes of my colleagues," he did, however, accept the mathematical equations of quantum mechanics, feeling however that it was an incomplete manifestation of the unified field theory, in which an objectively real description is possible. He continued the search throughout his life for a theory that would merge quantum phenomena with relativity. Unfortunately, he did not live long enough to see the emergence of superstrings, which most physicists today feel has that possibility.

Uncertainty of Music

It could be argued that music is also difficult or impossible to measure in its totality, unless you use waves as a yardstick. Once performed, where does it go? Does it affect space near and far? How wide is it? How deep? Perhaps the listeners carry its ineffable message to the universe. Some could argue that music is either particle or wave: particle in the sense of the note on the page and/or the sound wave as measured by its amplitude, frequency, and intensity. Music's mystery, too, is a match for quantum uncertainty. Both merit a deeper look.

The great musical classics of the Romantic, Classical, and Baroque periods are serving their historical purpose as compensatory balm for a world that has become increasingly painful in the twentieth and twenty-first centuries. It is difficult to predict the direction music and physics will take in the future; if anything is certain, it is uncertainty. We can only note the past and attempt to prognosticate from there. The crystal ball remains cloudy. The process of change and development in music, mirroring and often leading societal change, accelerates daily. Certainly, musicians, inspired by the turmoil of the times, will reflect, react to, and act upon those times.

Physicists keep coming up with papers suggesting a possible breakthrough from relativity and quantum mechanics. They too find little shelter in our stormy times. One can no longer look at it from the point of view of the solid, neat world of Newton and Descartes where what you see is what you get. In classical quantum mechanics, we can no longer measure a door or any other "solid" body or geometric figure quite precisely. Newton's easier world was wrapped up into a neat package; when you hit a wall, it was solid. You could measure its dimensions. But we can no longer count on that door being "solid." We know that the door is actually constructed of molecules, atoms, particles, and possibly single-dimension strings in a whirling-dervish dance. If we were small enough, we could walk right through the door or join its endless dance, according to the explorers of quantum mechanics. We are on the threshold of gaining from the collective progress and thinking of scientists, but instead of entering the door that they opened to a brave new world, we are rapidly descending to an unimaginable hell on earth. Outrageous thought: Can music save us?

40

How Is a Symphony Orchestra Like a Hologram?

In holographic photography, each individual part of the picture contains a condensed form of the whole. Thus the whole is in each part and, what is more obvious to our three-dimension-perceiving minds, each part is in the whole. To put it another way, the part has access to the whole and the whole has access to each part. Moreover, each part is in each other part.

While a hologram is a type of optical storage system, a symphony orchestra exists chiefly to make sounds. By some kind of mysterious magic, each player in the orchestra somehow knows what each other player has just done, what he is playing at the moment, and what he is going to play next. All this is on the "feeling" level as each player fits himself into the whole within the constantly changing dynamic. Moreover, the audience is also drawn into this dynamic realm. If you take the conductor or any single player away from the orchestra, that musician could probably re-create, in his mind, much of the music in which he was recently involved.

By the same token, if you take a holographic photo of any image—say a tree, person, or animal—and then cut out one section of it, say the tree's branch or a person's arm or brain, and then enlarge that part to the size of the whole, you achieve a picture of the whole animal, person, or tree. In other words, each individual part of the body or the tree contains the whole picture in condensed form. The same is true of DNA—the structure of the whole organism is encoded in each DNA molecule. Uncanny as this is, it is of a different order than our symphony player.

The principle inherent in music is not dependent on a camera or any other tool of modern technology. Long before cameras existed, Indra, a chief god of the Hindu religion, responsible for rain and thunderbolts, described the phenomenon in terms of the pearls said to hang over the gods' palace: "If you look at one pearl, you see all the others reflected in it." The author of the sutra explained: "In the same way, each object in the world is not merely of itself, but involves every other object and, in fact, is everything else."

Buddhist philosophy has long held that, as Fa-tsang explained in the seventh century, there is an ultimate connectedness and interpenetration of all things. He taught that the whole cosmos was implicit in each of its parts. He too compared the universe to a network of jewels, each one reflecting all the others. He understood that this explanation must also take into account the constant dynamic changes and interrelations among all things in the universe.

Jumping the centuries, seventeenth-century mathematician/philosopher Gottfried Wilhelm Leibniz, who is responsible for integral calculus, proposed that the universe is made up of fundamental units he called "monads," each reflecting the whole of the universe.

David Bohm

In the twentieth century, Einstein's theories and especially quantum theory gave the holographic principle a quantum leap. David Bohm, who won the Nobel Prize in physics for his work in subatomic physics and the "quantum potential," picked up the ball. Bohm came to the conclusion that physical entities that seemed to be separate and discrete in space and time were actually linked together in some fashion. His "explicate realm," the realm of separate things and events, is really part of an "implicate realm," undivided and whole. Like a hologram, his implicate whole (the entire orchestra) is simultaneously available to each explicate part (each individual player). He believed that the physical universe was like a vast hologram, with every part in the whole and the whole being in each part.

Karl Pribram, a Stanford University neurosurgeon, concluded that the brain functions much like a hologram. If one part of the brain is cut away, the rest of the brain takes over the functions and memories of the missing section. Doctors report many cases of people having a limb amputated and then feeling pain in that phantom limb. Somehow, according to Pribram, "the rest of the body encapsulates the memory system of the neurologic system of the missing part."

About a decade ago, physiologist Hugo Zuccarelli, improving on stereo recording, developed what he called "holophonic sound." Using Pribram's model of the brain, he developed a technique to create holograms of sound rather than light. As recording aficionados know, these recorded sounds are very realistic and three-dimensional.

Each player in a string quartet or a jazz group knows the precise timing, tuning, and "feel" of each other player. RCA Victor once asked me to record all four parts of a horn quartet. The procedure was for me to record the first horn part and then to play and record each subsequent part, placing one precisely on top of the other to attain perfect ensemble. Though the first horn part was played back to me each time through my earphones, it was a monumental problem to even start the second, third, and fourth parts exactly together. Though I knew the music very well indeed, and was "playing against myself," slight nuances kept tripping me up, take after take. Finally, I solved the problem when I decided to abandon "trying" (which causes tension) and chose instead to "feel," whereupon the remainder of the recording session went smoothly.

Alec Guiness, a great actor, obviously lived and automatically acted the hologram when he said, "I feel embarrassed when young actors talk about The Method. I just don't give it much of a thought."

This takes us back to the old subjectivity-versus-objectivity argument from my student days: Should we work on scales, arpeggios, and other technical problems, or "let the music happen," gaining technique by way of solving technical problems in the process of practicing the actual music. I have always believed that both were necessary, exercises and "the real thing." Scientists have admitted that some of their finest discoveries "just happened." What they do not say as often is that these epiphanies arose out of considerable prior work. In music that is called "practicing," all kinds of finger exercises (keyboard), bowing exercises (string instrument), and lip and breathing exercises (brass). We break down the whole into its parts before we can achieve the whole again. This is also true in painting. If you get close enough to a piece of art, you will see that it is broken up into small points and pieces. The same can be said in acting; instead of analyzing each small component, the actor "gets into the character" so that he or she "becomes" the person being impersonated.

My experiences in music led me early on to concur with Bohm that regardless of space or time, all events or objects, though seemingly separate, are really one and undivided. Music goes far in providing a profoundly spiritual (some might call it religious) experience of one's connection with the All (Nature or God, as you prefer). Philosophers of the past have recognized a phenomenon called "numanism," in which mystical experience is contingent on the tradition that contains the experience. It is an esoteric phenomenology in which the context of an event is crucial to understanding it. The basic experience of different people may be the same, though the stratum of experience may differ, as in the levels of

Samadhi described in Vedic literature. Bohm calls it an experience of the universal ground. He frequently mentions that the implicate realm is based on frequencies that, of course, are also the foundation of music.

To me, music is the ultimate religious experience. When my Catholic wife and I were interviewing and being interviewed by priests and rabbis (we wanted one of each for the ceremony), the priest, after questioning my wife while I waited outside, invited me in by myself. He began by asking me about my previous marriage, writing down my answers to this and other questions. He then asked, "What is your religion?"

"I was raised Orthodox Jewish."

Realizing that I was avoiding a direct answer, he then asked, "What religion do you *practice?*"

I thought for a moment and responded, "Music and people."

> *"Treat others as you yourself would be treated; this is the whole of the law, the rest is merely commentary."* —RABBI HILLEL

The truth is that I am comfortable with all religions, provided its followers practice the golden rule, though I am most at home with music, science, and the truly spiritual. The great religions knew the power of music and used it well to attract and indoctrinate their flocks. The oneness concept skirts in and around religion. Physicists such as Einstein, Heisenberg, Bohr, Schrödinger, and Eddington all played around the edges of a spiritual view of the world. The word "spirit" stems from "breath" or "esprit" or "the wind," all words that are integral to music and, if closely examined, are equally aligned with science.

Karl Pribram expressed it well: "Sensory reality occupies a domain beyond time and space where only frequencies exist. . . . Deep structure is essentially holographic." His theory has it that the brain structures related to sight, hearing, taste, smell, and touch can be measured by mathematical analysis of temporal and/or spatial frequencies. His theory takes in the whole spectrum of human consciousness, meaning, perception, intention, learning disorders, et cetera. He asserts that the most productive scientists "are as ready and as capable to defend spirit as data."

The Implications of Holography

The ancients may have intuited the conceptual framework of holography. There is an old Chinese saying: "Inside the poem, there is a painting; inside the painting, there is a poem." What a metaphor for life and for music! The Chinese reflected on the microcosmic to the macrocosmic, just as we might expand our thoughts from the individual to the family, to the community, to the nation, and then to the world.

The holographic implications for music were embraced by David Bohm, who agreed with me that his theories and holography might well be applied to music, wherein each note, phrase, and section is implicit in the ever dynamic whole, and the whole is implicit in each note or portion of music.

The idea of a hologram was first mathematically postulated by Dennis Gabor in 1947. A few years later Paul Nogier, M.D., in Lyon, France, discovered that there was an energetic system of the ear whose points were holographically related to the anatomy of the body. Acupuncturist and musician Dr. Ralph Alan Dale followed up on the work of Nogier, postulating the theory of "micro-acupuncture," giving it this name in 1975. He published extensive articles on micro acupuncture in 1976 in the *American Journal of Acupuncture*. His theory extended Nogier's paradigm of the ear to every part of the body, holding that "every part of the body manifested a hologrammatic energetic relationship with the anatomy of the body as well as with the full body pathways of acupuncture first developed by the ancient Chinese."

Dale has since published numerous articles and monographs, and has given many lectures and workshops on both Eastern and Western acupuncture. His workshops over many years have enabled doctors in New York and Florida to become certified to practice acupuncture. Part of his theory is that "the macro-micro correspondences in the energetic systems of the body are an expression of the hologrammatic nature of reality." I strongly suspect that Dale's musical background, as composer, as head of the music department at Hunter College, and as conductor of the C. W. Post College orchestra on Long Island, New York, played a significant role in his scientific accomplishments.

In 1965, after the discovery of the laser, Emmett Leith and Juris Upatnieks posed the idea that an actual hologram was created by the intersection of laser beams.

Just as a hologram can produce a three-dimensional visual image from a specially designed two-dimensional film, the renowned Dutch physicist Gerardus 't Hooft has suggested that all of the physical happenings we encounter may actually be encoded fully through equations defined in a lower-dimensional world. This may sound as strange as trying to draw portraits by viewing their shadows. Recently, work by Witten as well as Princeton physicists Steven Gubser, Igor Klebanov, and Alexander Polyakov has shown that, at least in certain cases, string theory embodies the holographic principle. According to Brian Greene in *The Elegant Universe,* some string theorists have suggested that fully understanding the holographic principle and its role in string theory may well lead to the *third* superstring revolution. I'm waiting for some enterprising researcher to explore the ramifications, macro and micro, of a single note of music, just as Lewis Thomas compared the whole earth to a single cell.

Nothing is absolutely fixed. Everything is constantly in a state of change and flux.

41

Peripheral Imagery

"Music creates order out of chaos; for rhythm imposes unanimity upon the divergent; melody imposes continuity upon the disjointed, and harmony imposes compatibility upon the incongruous."
—YEHUDI MENUHIN, conductor/violinist

My progress from being a "chair" player (a player in an orchestra) to a career as a concert artist and then a conductor was largely due to something that a teacher of mine, Murray Kahn, labeled "peripheral imagery." I had studied Konstantin Stanislavsky's *An Actor Prepares* and other works on acting as well as other approaches to the arts (such as taking dance classes) that made the procedure graphically clear. It may be described as the focusing of one's mind on the periphery of a note rather than the bull's-eye. This brings up the question of how big this bull's-eye is and how far from the center the outer orbit or periphery is.

But before getting into that thorny issue, here is one example of the technique I used for myself and with my students, many of whom subsequently joined major orchestras. As step one, I would ask the student to sit down. After a few seconds, I asked him or her to stand up again. The mystified student would wonder what I was getting at. I then asked the student to sit again, but this time, before standing up, to concentrate on *every single* muscular action required in the simple process of standing up. A few bewildered students would quickly rise, but the more thoughtful students remained sitting, uncertain of what to do. The most intelligent would try to begin the impossible task I assigned—remaining still. Then, relieving their misery, I would explain that every single physical act—in this case the simple motion of standing from a sitting position—requires untold millions of coordinated muscular actions triggered by neurological messages in a tremendously complicated procedure, far more intricate than the central telephone switchboard of a large city. Yet all humans, no matter the level of their education or talent, manage all this simply by telling themselves to stand (or to take any other simple action). Without delay, all the billions of wheels and wheels within wheels automatically mobilize to set in motion what is required to accomplish this relatively ordinary goal of standing up. When I proceeded into the intricate motions of playing the French horn and the attendant task of getting the *orbicularis* (lip) muscles to vibrate at precisely the frequency necessary to discriminate between one note and another while trying to concentrate on the muscular dynamics (an impossible task), the idea began to strike home. All we have to do is think of the note we want to play and how we want it to sound, and our incredible bodies find a way to obey our command.

When you think of the accomplishments of a virtuoso on the violin or the organ, the concept is still more astounding. A violinist places a wooden box (a very expensive box indeed!) under his chin in a most uncomfortable and unnatural position, holds it there by pressing chin against violin against shoulder, then takes a stick (called a "bow"—also expensive) and draws it across four strings that are attached to the box in such a way that these strings can maintain their tension at predetermined frequencies (the notes G, D, A, and E) and can be stretched or relaxed to precise frequencies to adjust the tuning. The four strings are then "stopped" by the left-hand fingers against our wooden box in precisely exact positions while these fingers shift (slowly or very rapidly) to other precise positions as called for in the music. The player must hold the left hand in such a position as to be able to move the fingers to stop the strings up and down the four strings' lengths, sometimes with great rapidity, vibrating the hand (appropriately called a "vibrato") so as to achieve the exact required wave frequencies to make the desired and constantly changing sound effects.

But that is not all. Meanwhile, back at the right arm, the bow, also held in an uncomfortable position in front of the body, must be drawn back and forth to match the music's requirements of sound and duration. All this is done in midair, floating the arms, hands, and fingers in front of the player's body in a very unstable position. With all this mélange of putting arms, body, and fingers to working together, the player must make music, by varying the length, speed, and intensity of the bow's pressure on the string(s).

When Yehudi Menuhin, the great violinist, also an intellectual in the best sense, reached his mid-teens, he began to analyze his playing down to the smallest detail. This fragmentary approach almost crippled his playing (at least from his usual superb perfection) for a time. An extraordinary thinker in addition to being a wonderful player, touted all over the world as a wunderkind as a boy, he paid the price in his adolescence for concentrating on the minute motions of his bodily parts. He needed to get off the bull's-eye and return to the peripheral approach by aiming at the larger picture and "allowing" his bodily memories (achieved by years of practice and performance) to take over. We humans are truly miracles of coordination, muscular and otherwise!

The peripheral-imagery concept is commonly used by the better actors and dancers as well as by yoga and fitness trainers, who urge their students to imagine one thing or another to enhance their art by releasing them from the stressful and rigid notion that they must "hit the bull's-eye" in terms of direct control of muscles that are not subject to such isolated control.

In talking about his theories of the explicate and implicate orders, David Bohm brings up movement and music: Our most immediate experience of the implicate order is movement itself. We do not actually know how we manage to move. We have a wish to go somewhere, and the imagination displays the activity we want to accomplish, but it doesn't tell us how to achieve it; this remains mysterious to us. In light of that display of imagination, we manage to move, but we could not really describe how we did it. There is evidence that there is an intelligent internal life-energy that can perform action once something is displayed to it.

There was a case in which one end of a thin wire was attached to a single nerve in a person's hand and the other end was connected to a loudspeaker, which displayed an action of that nerve by a click. Once the individual heard the click, he could reproduce it at will, and although he could not say how, he eventually learned to play a rhythm. The point is that when activity is displayed, we can bring some order to it, but without display we can do nothing about it. Indeed, all of our consciousness is a display of past information fused with present sensory data. That display is the unfolded world—everything that we know about the explicate order.

In music, and in visual and other sensory experience, the implicate order is primary in the sense that flowing movement is experienced before we analyze it into the elements that express that movement. We are grateful that violinists carry on, making music with nary a thought about the theories behind their music-making!

42

The Mystery of Coincidence

coincidence *"an accidental and remarkable occurrence of events or ideas at the same time, suggesting but lacking causal relationship"*

synchronicity *"happening at the same time; simultaneous occurrence . . . having the same rate and phase, as vibrations"*
> —Webster's New World Dictionary

We have all experienced coincidences. A person whom we haven't seen or thought of for a long time appears in our dreams or suddenly passes through our minds for no apparent reason or causal connection. We soon get a phone call from that person or hear that this person has died. We are in a strange city, and we run across somebody whom we know very well. In an out-of-our-way place, perhaps another country, we hear somebody yelling our name. We are shopping in a supermarket and practically bump into a person we would least expect to see there. "What a coincidence," we exclaim. "I was just thinking of you!"

An extremely unlikely coincidence in our planetary system is the size as we view it on earth of the sun in relation to the moon; both look approximately the same size. Given the immensity of the universe and the enormous variety of sizes of heavenly bodies, the mathematical probability that two heavenly bodies would be so similar in apparent size is extremely remote (one in a billion? one in a trillion?). The relative sizes to us of the sun and moon have further significance. If it weren't for eclipses, which can only be seen with some exactitude by the very small apparent difference between the breadth of the sun and that of the moon, one wonders if there would have been as much speculation and interest by people throughout history about the cosmological universe. When one figures that the sun is enormously greater in size than the moon, and it is about four hundred times farther from earth than the moon, one is even more amazed at this coincidence.

The eclipse that occurred in 585 B.C.E. was so startling that it ended the five-year-old Peloponnesian War. The Athenian soldiers were so terrified by the eclipse that they reluctantly left Syracuse. They thought it was a bad omen. Their delay ensured the success of the enemy.

Christopher Columbus

Christopher Columbus exploited his foreknowledge of an eclipse when his men were facing starvation. He told the natives that his god was displeased with them for not assisting with food. As the earth's shadow began to fall across the moon, the natives quickly agreed so long as he brought back the moon. Columbus told them he would persuade his god to restore this light of the heavens. On the return of the moon, Columbus had no further food problems and his men were able to return to Spain.

Some decades ago, I took a three-day course in Sylva Mind Control. We learned relaxation techniques and other uses of the mind. For our graduation exercise we were partnered with another student. One of us would think of a friend or family member, and tell the gender, where this person lived, and his or her approximate age. From this limited information about a total stranger, the other person would enter a meditative state, try to "visualize" this person, and tell what he saw. In three different cases, I was somehow able to relay some information about their health or what general illness they were suffering from. Though the accuracy of my "guesses" surprised me, I simply ascribed my success to a kind of coincidence or "reading of the mind" of my partner in crime. A few other graduates had similar success. This experience long predated my knowing about quantum mechanics and the communication that goes on between particles, but it helped open my mind to human possibilities, even in our present era of closed-minded, Newtonian desensitization.

Coincidences are often given a religious twist. God somehow has ordained, knows about, and controls every little detail of our lives and all the billions of human beings on earth. He then manipulates these lives according to how well or badly behaved we are. Some freethinkers ask what reason God could have for killing babies and innocent people.

But even some hard-nosed scientists, such as biologists and psychologists Drs. Eberhard and Phyllis Kronhausen, tell of marginal experiences. One time they were returning to their home in Paris after a lengthy stay in other European countries, only to find a patient who had no idea of their schedule waiting to see them at the railway station. They tell of another time when they arrived at their hotel in Phnom Penh to find a group of students who had been waiting twelve hours in their hotel to see them. In both cases, without any foreknowledge, the greeters somehow "knew" of the imminent arrival. The good doctors became firm believers in telepathy.

The old argument of divine design versus evolution has been resurrected in the last decade or two. There is demand that creationism be taught in the public schools, and a compromise is sometimes made to teach it alongside Darwinism. Many of the religious right harbor a profound distrust of science and scientists. There is a strong connection between the revival of antievolutionism, the attack on separation of church and state, the decline of American science, and the ascendancy of the religious right in government and media. However, most scientists maintain their confidence in a "prove it to me" science, always a difficult stance for the Great Designer school. God seems to be on the sides of the most contradictory isms, religions, political parties, and individuals. How He/She manages to keep up with all this must keep whole armies of heavenly clerks and secretaries busy sorting it all out.

Perhaps quantum mechanics or a theory of everything will one day provide answers to the puzzles of theology and science.

What About Choice; Do We Have It?

Every living thing influences, however indirectly, every other living thing. That principle holds true for the past as it relates to the present and possibly even the future. The effect of the past on the present is not hard to grasp. Much more difficult to ascertain is how both past and present might affect the future. This requires a science of society and the earth, something we have yet to develop. We neglect this science research at our peril.

Perhaps the future may not be so hard to imagine after all. We all know of the changes in the environment resulting from the burning of fossil fuels. And we may wonder about how a huge tax cut and rebates for the wealthiest 1 percent of people in the nation affect present and future allotments for education, health benefits, or Social Security. So today's decisions, taken out of context, without considering the flow of events before and the emerging possible results, are likely to be wrong or disastrous. Since change is inevitable anyway, decision-makers must focus on process. Then the big question is: What kind of change? That's where choice comes in, both individual and group choice.

The mind is involved in an intersubjective process of communicative exchange. No such thing as a separate or isolated mind exists. Physicists attend frequent conferences to get an update of what other physicists are doing. Musicians virtually enter the hearing and thinking process of every other player in their group. We've all marveled when we watch a string quartet simultaneously start a passage without any noticeable signal, or when each player of a jazz group seems to know exactly when to make a modulation. It's like a school of fish or a flock of birds, where all of the members seem to know the exact moment when to change direction.

Some Hidden Signal?

As the group increases in size, the mystery increases. In a symphony orchestra, the structure of interrelationships becomes sufficiently complex so that we need a conductor to keep things together. We can extend the principle to a tribe or a nation, and eventually to the planet or the universe. In a one-to-one relationship, the empathy is remarkable, to which lovers and married couples will testify. Without words, they can sense, anticipate, and react to each nuance of feeling from the other. They know what the other is thinking.

The whole orchestra is indeed greater than the sum of its parts; each part relates to every other part and to the whole, in a constant dynamic of change. Each part is a hologram of the whole and the whole a hologram of each part. If you replace the word "part" with the word "musician," the symphony orchestra remains a convincing example of this holographic principle. Can we extend this principle to society as a whole?

43

Physicists and the Music of Brain Waves

The bonds in physics research are continually being broken. Arising out of string theory, we have the "many universes" theory invading the mass media as a topic for everyday discussion by the lay public. Distinguished physicists appear like political pundits on National Public Radio. The search for the secrets of the origin and destiny of our and other universes continues, despite our Newtonian, three-dimensional addiction. Research on the workings of the brain has also proliferated. Although curiosity and speculation about consciousness and the workings of the brain have fascinated philosophers and scientists for thousands of years, it is only in the past few decades that we have the technology to "see" into the brain and measure reactions. Many universities and research centers have been producing information and speculations backed by magnificent machinery undreamed of when sages, writing in the Hindu Upanishads 5,000 years ago, claimed that all was consciousness. How uncanny that the conclusions of ancient sages often parallel quantum theory!

In 2002 I had the opportunity to visit one of today's excellent facilities at Florida Atlantic University. Dr. Ed Large had invited me to meet the eminent Dr. J. A. Scott Kelso, head of the department at the Center for Complex Systems and Brain Science Research at the university. Ed, Dr. Kelso, and their colleagues were anxious to further explore the musical aspect of their research. Apart from dozens of sophisticated computers, there were MRI (magnetic resonance imaging) and FMRI (functional magnetic resonance imaging) machines and other complicated machinery in specially designed rooms.

This visit triggered my decision to delve into other work in the field. Soon I discovered a gold mine of material, much of which is germane to music and physics. Following are a few examples.

The October 1973 issue of *Scientific American* published a paper entitled "Auditory Beats in the Brain." Dr. Gerald Oster of Mount Sinai Medical Center described how pulsations called "binaural beats" occur in the brain. When tones of slightly different frequencies (say 20 Hz apart) are presented separately in each ear, the entire brain begins to resonate to the frequency that is equal to the difference. This phenomenon is called "entrainment of brain-wave patterns."

Almost thirty years later, journalist Jim Robbins published a book, *A Symphony in the Brain: The Evolution of the New Brain Wave Biofeedback* (Atlantic Monthly Press), wherein he chronicles his battles with chronic fatigue syndrome. He had tried traditional therapies to no avail. He had his scalp hooked up to a computer display via electroencephalogram (EEG) sensors and began a session of brain training. "After a half hour my mind was tired, my thoughts muddled. But an hour or so after I finished, I experienced what is known as the clean windshield effect. The world looked sharp and crystalline, and I had a quiet, energetic feeling that lasted a couple of hours. It was the first time I felt like that in years."

Biofeedback is a technique of seeking to control certain emotional states, such as anxiety or depression, by training oneself, with the aid of electronic devices, to modify and control autonomic body functions such as blood pressure, heartbeat, or bodily temperature. About twenty-five years ago, Jerelle Kraus, an editor of *The New York Times* op-ed page, was interviewing me for a story. In the course of our discussion, she asked me about some of the avant-garde areas I had been exploring. She asked me to arrange an interview with an Indian guru. He agreed to see us at his headquarters in New Jersey's rolling hills. In the course of that interview, Ms. Kraus asked him if he would consider demonstrating some of his powers. He reluctantly agreed, maintaining that these "tricks" are only superficial to the more basic understanding of his yoga. He first asked us to feel the normal temperature of his hand, and then he spent about a minute in quiet meditation. When we felt his hand again it was ice cold! He went back into another minute of meditation, and we then felt his *extremely warm* hand. All this was in the course of less than two minutes! Thus we served as feedback machines and were amply convinced about the power of the mind on the body.

Biofeedback machines connect the body's natural rhythms, such as its brain waves, to monitors, which display amplified electrical frequencies of the brain or unconscious occurrences such as blood pressure and heart and lung action. By watching these on a computer screen, participants are able to influence their physical and mental well-being. In this kind of neuro-feedback, patients are trained to function with brain frequencies they do not generally use (except in sleep). This exercise apparently strengthens the brain and also functions as self-therapy.

Researchers have also delved into the use of biofeedback to determine its value in treating such conditions as attention deficit disorder, autism, epilepsy, head injuries, learning disabilities, depression, addictions, and post-traumatic stress disorder. The medical profession is generally dismissive of the therapy. But neuro-feedback, though it may not be miracle or panacea, works. So we can say that it is science.

Dr. Arthur Hastings, in a paper entitled "Tests of the Sleep Induction Technique," describes the effects of subjects listening to a cassette tape specially engineered to create binaural beats in the brain. In one case, the sounds on the tape were designed to slow the brain-wave patterns, first from a normal, waking brain-wave pattern (beta) to a slower pattern (alpha), then to a brain-wave pattern of dreaming sleep (theta), and finally to the brain-wave pattern of dreamless sleep (delta). In the same vein, Dr. Bill Schul writes, "The FFR [frequency following response] evokes physiological and mental states in direct relationship to the original stimulus. With the availability of this tool it becomes possible to develop and hold the subject into any of the various stages of sleep, from light Alpha relaxation through Theta into Delta and in REM (dreaming)."

Apparently, as the brain-wave patterns are slowed from beta to alpha to theta to delta, there is a corresponding increase in balance between the two hemispheres of the brain. Some scientists have noted that these slower brain-wave patterns are accompanied by deep tranquility, flashes of creative insight, euphoria, intensely focused attention, and enhanced learning abilities.

Dr. Lester Fehmi, director of the Princeton Biofeedback Research Institute, finds that "hemispheric synchronization" (sounds like a symphony orchestra to me) represents "the maximum efficiency of information transport through the whole brain" and "is correlated experientially with a union of experience. Instead of feeling separate and narrowly focused, you tend to feel more unified with the experience. . . . The scope of your awareness is widened a great deal so that you're including many more experiences at the same time. There's a whole-brain sensory integration going on, and it's as if you . . . function more intuitively."

Speech-language pathologist Suzanne Evans Morris, Ph.D., says: "The introduction of theta signals into the learning environment allows for a broader and deeper processing of the information and increases focus of attention, creating a mental set of open receptivity. Children exhibited improved focus of attention and a greater openness and enthusiasm for learning." She advances this theme: "Research supports the theory that different frequencies presented to each ear through stereo headphones create a difference tone (binaural beat) as the brain puts together the two tones it actually hears."

Binaural beats have been used in the classroom to enhance learning ability. Teachers in the Tacoma, Washington, public schools, under the direction of psychologist Devon Edrington, used audiotapes that incorporated binaural-beat technology to influence learning ability of students. They found that the students who studied, were taught, and took tests while the tapes were playing did significantly better than a control group not using the tapes.

Many studies have proven overwhelmingly that students who study music improve their abilities in other classes, and later, in all professions. Music training in the schools results in motivated students who improve in their nonmusical subjects as well.

Memory

With the aging population and the growth in Alzheimer's and other neurological diseases, we are besieged by literature on memory. Animal research on memory found that, in all cases, the more theta waves appearing in an animal's EEG after a training session, the more it remembered.

Neurophysiologist Dr. Gary Lynch at the University of California discovered that "the key to LTP [long-term potentiation, i.e., electrical and chemical changes in the neurons associated with memory that allow for the storing of incoming information] is the theta brain wave pattern, the magic rhythm—the natural rhythm of the hippocampus, that part of the brain essential for the formation and storage of new memories and the recall of old ones."

These mental/emotional changes probably take place, researchers believe, as different wave patterns are linked to the production in the brain of various neurochemicals associated with relaxation and stress release, increased learning and creativity, and memory. These neurochemicals include beta-endorphins, acetylcholine, vasopressin, and serotonin.

Scientists now believe that the moment when learning takes place—the "aha" moment—is when a particular reality has been selected and filtered by our endorphins and is suddenly apprehended by the brain in such a way that something new is learned, this learning being rewarded by a flood of endorphins along our pleasure pathways, so reports William Harris, the director of the Centerpointe Research Institute.

Alpha and theta patterns are also correlated to the "relaxation response"—the opposite of the "fight-or-flight response." The latter response takes blood away from the brain and toward the periphery of the body, floods the bloodstream with sugar, and increases heart rate, blood pressure, and breathing in order to prepare one for "fight" or "flight." In this state, learning ability, as well as other mental functions, including problem-solving and reasoning, are inhibited. The fight-or-flight response is accompanied by low-amplitude, high-frequency beta brain-wave patterns, while the relaxation response is accompanied by high-amplitude, low-frequency alpha and theta rhythms.

Dr. Vincent Giampapa, M.D., of Longevity Institute International and vice president of Anti-Aging Medicine, finds that placing a listener in the alpha, theta, and delta brain-wave patterns dramatically affects production of three hormones related to longevity: cortisol, DHEA, and melatonin.

Dr. Benjamin Libet wired up his subjects with electrodes that measure brain activity, and seated them in full view of a rapidly rotating clock hand that enabled them to note exactly when they "ordered" their finger to flex. Libet could then mark three events in time: the onset of increased brain activity recorded by the electrodes, the flexing of the finger, and the point at which each subject had consciously willed his finger to flex.

What Libet found was that, in each instance, a flurry of brain activity took place a fraction of a second before the "order" to flex the finger was dispatched by the conscious mind. "In other words," says Libet, "their neurons were firing a third of a second before they were even conscious of the desire to act. Hence, it appeared the brain had begun preparing for movement long before the mind had 'decided' to do anything."

"The illusion of conscious control is maintained," Libet notes, "because another mechanism in the brain delays the sensation of the finger moving, so that the conscious mind continues to think that it has first decreed the action, then felt the muscles act. Actually, by the time the mind orders the finger to flex, the impulse has already been dispatched.

All the mind gets is a last-minute opportunity to veto the decision: I can stop my finger from flexing by sending an intercept command that overtakes and interrupts the original command and thus keeps my finger immobile." (This is what happens when you reach for a plate in the kitchen and then stop yourself upon remembering that the plate is hot.) Dr. Libet continues:

> The mind is thus permitted to sustain the flattering illusion that it controls the game. In actuality it is playing catch-up ball.
>
> This made it possible, by studying split-brain patients, to identify certain functions as localized in one or the other hemisphere. Language, for instance, turned out to be a function primarily of the left brain. When a word is flashed to the right hemisphere of a split-brain patient, she cannot tell the researcher what the word was. The left brain, which handles speech, does not know what to say, because it has not seen the word. The right brain knows, but cannot speak. It can, however, answer questions in other ways. In an experiment, a subject's right brain was shown a picture of an apple; he could not say what he had been shown, but when his left hand (the hand controlled by the right brain) was given several hidden objects to choose from, it picked the apple.
>
> Generalizations about the proclivities of the right and left cerebral hemispheres . . . were sometimes put to rather facile uses. Writers were declared to be "left-brain" types; painters to be "right-brain" dominated. Golfers and tennis players were trained to engage their right-brain functions in order to play more naturally and gracefully. School administrators endeavored to address the supposedly neglected right brain by putting more stress on arts and crafts.

The implications of localized brain functions can also help us understand the unity of mind. Each side of the brain is made up of many modules that operate more or less independently from the other side, and the function of the mind is not so much to tell the different units what to do as to try to make some coherent sense out of what they already have chosen to do.

The mind may rule the self, but it is a constitutional monarch. Presented with decisions already made elsewhere in the brain, it must try somehow to put on a good show of having them add up to some coordinated, sensible pattern.

The brain is analogous to a computer in that it disguises a multiplicity of operations behind a unified facade. The computer on which I am typing this sentence is busy doing many things at once—one part of it is keeping track of time, another is searching sectors in one of its disk drives, another is moving blocks of data here and there in its memory— but the image it paints on the screen is coherent and unitary, like the picture presented to the mind by the brain. At the moment, the computer image replicates black letters inked on white paper. If I press a few keys to access another program, the image could change to replicate a chessboard, the stars over Padua on a summer night in the year 1692, or an air battle over the Pacific in 1942. In every instance, the unified image is a scrim, presented by a program that in turn interfaces with other programs. The brain similarly renders the multiple functions of its several programs into a pleasing if illusory unity.

William Harris, who provided much of the foregoing information, wraps up: "Clearly we are on the frontier of a marvelous new field with untold possibilities. The ability to map and entrain brain waves and the states they represent, gives us a powerful new tool to effect human change and growth. It has been shown that induced brain wave states can effect super learning, increased creativity, sleep induction, pain control, behavior modification, focusing of attention, relief from stress, increased longevity and slowing of the aging process, increased memory and dramatic improvements in mental and emotional health."

All of the above assumes a subject capable of reacting "normally" to the electronic stimuli and absorbing their effects. This excludes those who are severely damaged. The research and experimentation for these invididuals would have to take other paths. For example, I've witnessed some miraculous results of music therapy utilized with autistic children who were otherwise totally unreachable. The rate of autism in children has tripled in fifteen years. What are we to think of the three-year-old blind, autistic Derrick, who came to life, playing back whole songs and symphonic excerpts on the piano after one hearing? Another autistic child came out of her lost world to ask, "Are we going to play the violin today?" Eventually she learned social skills, and now sings in her school chorus. Teacher/choral leader Ellen Winnick reports many such experiences. Art therapy also has a beneficial outlet, especially with adults and older people. A study in three cities found that the health of people sixty-five and over improved as a result of participation in music and the arts.

But Is Computer Art Really Art?

Computer art threatens to overturn centuries of reverence for the concept of an "original" work of art. Gary Glenn, for one, attacks it: "In computer art there is no direct encounter with the materials. Traditional materials show paintbrush strokes or chisel marks. Perhaps photo art is a distant cousin of computer art though in fine photographs there is a record of the artist's presence in the dark room. The photographer's eye is closer to his soul than his fingers on keys." Computer-generated art is characterized by objectivity. Humanness is drained from it. Nonetheless, computer-generated art appears on television and in print media, and is displayed in the world's most famous galleries. Multimillion-dollar movies are built around the effects that only computers can create. Is the hemorrhaging of humanity in our age reflected in computer art as it is in electronic music?

Like painting, music is universal among human cultures and can be traced back to the dawn of recorded human history. Electronic music, regardless of all kinds of advances in the expensive gadgetry needed to produce this music, still lacks the immediacy, the presence, and the direct contact of a musician playing an acoustic instrument. Electronic artists could argue that one of music's definitions is that it consists of organized patterns. This is true, but the difference, I think, lies in whether these patterns are organic (human) or inorganic (electric).

The question of patterns has manifestations that are especially interesting. In the Middle East, artistic representation of living things is forbidden, yet the decorative designs in this region in the form of mosaics and friezes, laboriously created, are intrinsically human. The Islamic tradition is particularly notable in this regard because, though the Koran forbids the representation of living things for decorative purposes, the Arabs developed a whole spectrum of complexity that geometry allows, both on flat and curved surfaces. Geometry has also been a necessary element in the development of and advances in mathematics and physics. Artist M. C. Escher created wonderful, paradoxical art based on creative geometry that has led to new thinking in geometry as well as in art.

Music Patterns and Symmetry

The criterion to which most theories in physics are held is symmetry. As we have seen, most music, even without pre-thinking by composers, achieves symmetry. One can examine almost any song or symphony and discover symmetry. This also holds true for folk music, popular, rock, rap, and jazz. The exceptions don't stay alive very long.

Within the symmetrical are patterns: patterns of behavior (animal and plant) and patterns of structure, sound, light, and time. Mathematics is indispensable in science because math is the study of all possible patterns. We ourselves are prime examples of organized complexity as part of the patterns in the ever evolving universe.

Modern wave mechanics corresponds to ancient harmonic and geometric ideas of universal order, going back to Pythagoras, Aristotle, Plato, and others. Bertrand Russell, in his *Analysis of Matter,* says: "What we perceive as various qualities of matter, are actually differences in periodicity." In other words, patterns of music.

Pythagoras established the relationship between number ratios and sound frequencies. In ancient times, the laws of elementary harmonics were thought to define the relationship between the movements in the heavens and developments on earth. Geometry, or "measure of the earth," is the study that measures spatial order, beginning with the fundamental circle, triangle, and square. Plato considered geometry and number as the most essential and "ideal" philosophical language.

The geometric helix form in the molecules of DNA is the carrier of continuity, responsible for DNA's replicating ability. One might say that the architecture of our bodies is determined by a world of form and geometry. The molecules of plants are also arranged in complex geometric patterns, often in patterns of twelve, corresponding with the twelve tones of the musical scale.

It has been said that all our sense organs function in response to proportional or geometric differences, depending on the stimuli they receive. Thus, proportional differences of the frequencies of waves determine a sound we hear, the scent of a flower, the sight of a loved one, or the touch of a kiss. It is notable that the form of a horse's bridal has led modern physicists to overturn some previous Newtonian "laws," such as "parallel lines can never meet." This form has also enabled physicists to imagine dimensions beyond the four we are accustomed to in our "real" world. Pattern, geometry, number, and proportion connect with order, symmetry, form, and continuity to make up music.

The Tempered Scale

Musical patterns break down to notes and their relations to one another and with time, space, movement, and frequencies. The tempered scale is an accommodation for the piano keyboard so that the same black key may be employed for both, let's say, a D-sharp and an E-flat (or any other "enharmonic" differences, such as A-sharp and B-flat). The actual difference between these enharmonic notes is approximately a ninth of a tone. Since this difference is so small, the ordinary ear accommodates to this, making an automatic "correction" to the "mistuning." However, a sensitive musician with a refined ear can detect this mistuning. A good violinist will automatically make the correction, whereas the pianist, tied to the tempered scale, doesn't have this luxury.

When I was a horn player, I was quite aware of these differences and would "bend" the tone with my lips to make the accommodation quite satisfactorily. Other fine musicians do the same, even with some wind instruments that do not easily bend tones.

44

Mathematics: Skip This Chapter

(But I Urge the Reader Not to Take My Advice)

*"The enormous usefulness of mathematics and natural
science is something bordering on the mysterious.
There is no rational explanation for it."*
  —Eugene Wigner, "The Unreasonable Effectiveness
  of Mathematics in the Natural Sciences"

Mathematician J. W. N. Sullivan wrote, "Since the
primary object of scientific theory is to express
the harmonies which are found to exist in nature, we see
at once that these theories must have an aesthetic value.
The measure of the success of the scientific theory is in
fact a measure of its aesthetic value since it is a measure
of the extent to which it has introduced harmony in
what was before chaos. The measure in which science
falls short of art is the measure in which it is incomplete
as science."

Music and math are closely related, but it is a rare
musician who really knows or appreciates math, whereas
I find that many mathematicians love music and could
easily explain the previous chapter. Musician Einstein
apparently shared a bit of the sentiment of musicians
when he said (in his physicist role): "The use of mathe-

matical symbols cannot be completely avoided." The musician part of his brain was probably coequal with the physicist portion. Though it's impossible to follow all of his thought processes, it is fascinating to read about them from 1908 to 1911 and from 1912 to 1914 in working out the impossibly complex considerations that he was dealing with. He talks about the years of "anxious searching in the dark with their intense longing, and the final emergence into the light."

Math, of course, is a science of numbers. The number zero, if it is indeed a number, has transformed mathematics, commerce, physics, and our very civilization. Without the zero, how would we count to ten, except on our toes or fingers? And counting to a hundred, a thousand, or the geometric numbers that physics deals with would be quite a task indeed. By simply adding a zero, we add a power of ten. The word "zero" seems to have its origins in the Sanskrit *sunya,* meaning void. It may have been the Arabic *sifr,* which in Italy turned into *zeiro.* For India, the spawning ground of many great philosophies, zero was important to the Hindu concept of the void, and to this day, this "nothingness" has fascinated scientists, scholars, and seekers in both East and West. The concept has inspired at least two books, *Nothing That Is: The Natural History of Zero* by Robert Kaplan and *Zero: The Biography of a Dangerous Idea* by Charles Seife.

Calculus is a shorthand language in mathematics, extremely useful for physicists and scientists, just as letters and words are the shorthand for communicating ideas, and music notes are the shorthand language of music. I doubt whether Einstein, Shakespeare, or Beethoven would have been able to realize their genius without these symbols. Equations and musical notes are simply vehicles to understand, formulate, and then carry forward ideas and systems of thought, to us and to others.

Calculus is a highly systematic method of treating problems by a special system of algebraic notation. It is useful for equations such as $E=mc^2$, which cannot be reduced further in any consistent way. Many philosophers, mathematicians, and scientists have struggled for centuries to elaborate a thoroughly consistent and provable map for their theories. For instance, they have struggled to prove the consistency of a system such as in Sir Isaac Newton's *Philosophiae Naturalis Principia Mathematica,* the famous scientific tome that won acceptance by its beautiful simplicity as practical mathematics rather than as abstract philosophy.

An attempt was made to find a vocabulary and an adequate, logical apparatus to express the whole of arithmetic rather than just a fragment. All the attempts to construct such a proof were not successful. Finally, the publication of Kurt Gödel's paper of 1931 showed at last that all such efforts must fail, concluding that it is impossible to give a meta-mathematical proof of the consistency of a system comprehensive enough to contain the whole of arithmetic unless the proof itself employs rules of inference.

Aristotle claimed that two parallel lines could never meet. For many centuries, mathematicians and philosophers believed this to be absolute truth. Newton put it into one of his "laws." This "law" was later easily disproved by use of the planet or a circle. Gödel also arrived at the conclusion that there is a fundamental limitation in the power of the axiomatic method. He showed that *Principia,* or any other system from which arithmetic can be developed, is essentially incomplete.

Mathematics abounds in general statements to which no exceptions have been found but that, thus far at least, have also thwarted all attempts at proof. A classical example, Goldbach's theorem, states that every even number is the sum of two primes. No even number has ever been found that is not the sum of two primes. Yet no one has succeeded in finding a proof that Goldbach's conjecture applies without exception to all even numbers.

Gödel showed that even if the supposition is correct, there would still be no final cure for the difficulty. He demonstrated that there would always be further arithmetical truths that are not formally derivable from the augmented set (from a book called *Gödel's Proof* by Ernst Nagel and James R. Newman). Gödel did show that metamathematical statements about a formalized arithmetical calculus could indeed be represented by arithmetical formulas within the calculus. In other words, it can be represented within the parameters of an accepted system. He showed that axioms are inherently incomplete. Equations and formulas, daunting to amateurs such as myself and probably most of my readers, can be skipped because they are simply shorthand for discovering truths within an accepted system. That is why some of the greatest scientists of our time still conjecture that there may be other universes aside from our own. Other truths and conditions may alter these seemingly absolute truths. Hence the word "theory" connected to all of the great insights and discoveries. These theories work very well within the restricted parameters in which we spend our lives. Theories such as general relativity and quantum mechanics work very well indeed within our framework of knowledge and within our reality. Superstrings seem to compute by a series of weeding out the non-useable and maintaining what works mathematically.

Numbers: How to Win
Life's Lottery

Music can be said to be a system of numbers. Like the "1,0,1" configurations in computers, from which the infinite variety of configurations permits an entire Library of Congress on the small card in our computers, so too do endless configurations of waves permit an infinite number of musical compositions to be on your computer and in the world at large. Numbers are integral to music. Metaphors help us to better understand the significance of numbers.

The number 1 is a metaphor for unity, identity, and the experience of "oneness." It is altogether different from other numbers in that all other numbers divide entities or processes into their substages. The number 1 has no divisions, no separations, and no alienations. It can remind us that humanity is indivisible, that all living forms and the planet we occupy are integral, and that our cells, tissues, and energies are simply multiples of a single reality. The universe itself is One.

The number 2 reminds us that everything in life tends to express itself in terms of polarization. Chinese philosophy and medicine as well as

Hegelian philosophy present endless examples of this antithetical relationship, between yin and yang, thesis and antithesis.

The number 3 is a metaphor for outcomes. Hegel, in his thesis-antithesis-synthesis system, expressed this window to the nature of all processes. The relation of thesis and antithesis is the very same one as the Chinese observed in their metaphor of 2: yin and yang. However, it is only the metaphor of 3 that invites us to ask the key question: What happens as a result of the polarization of these two opposites? What is the outcome of the reaction between yin and yang? In Chinese medical terms, we might say that biphasics (yin and yang) is about diagnosis, while triphasics (thesis-antithesis-synthesis) is about the transformations induced by treatment. Triphasics is also a metaphor for prophylactics, the maintenance of the healthy balance between yin and yang. It's the chief province of the master physician and the model patient, both of whom are always supporting the body's defense system in their efforts to maintain homeostasis. That is to say, since there's nothing in life that is static, yin and yang can never be expected to remain in balance. New imbalances are always being created by the very fact that life means motion and change. The master physician and the model patient understand how to intervene in these ever reproducing imbalances before they turn into illnesses.

Processes always have outcomes. Yin and yang are idealizations of these polar opposites, which without their outcomes (syntheses), are only abstractions. There cannot be any yin and yang in real life without their being in motion, without their being in constant interaction, without a continuous generation of outcomes. In this sense, the metaphor of 2 is a frozen moment, an ideal balance, a static, which the mind can invent but which can never exist in experience or in the nature of processes.

Numbers in History

Greek philosophy of the sixth century B.C.E. (Pythagoras, for example) and the fifth century B.C.E., (epitomized by Plato) were focused on ideals, abstractions, and permanencies. Greek interest in the metaphor of 4 precedes the sixth century B.C.E., probably having its origins in Homeric times (eighth century B.C.E.) and the beginnings of Greek civilization. How may we account for the early Greeks' preference for a metaphor of the four elements and developing melodic structures built on the tetrachord (four tones)? And why, in contradistinction to the ancient Greeks, did the Chinese and every other ancient civilization develop an interest in five as an organizing principle, creating their melodic structures pentatonically (five tones)?

The questions have yet to be more fully answered as to the concatenation of musical forms with cultural, economic, and social structures.

45

More Connections

The connections between music, notes, phrases, sections, and movements of a composition are what constitute the integration, the "oneness" of a work of music or of any artwork. One can scramble these and concentrate on bits, but these do not a work of art make. Nor does such a separation of elements make up a viable scientific theory. Despite the current popularity of science and science fiction, society, racing through transitional stages, has been glacially slow in making the connections and differences between the old ways of Newtonian mechanism and a more timely, holistic approach congruent with the new physics. Nations and individuals are still behaving and functioning according to an obsolete worldview. Newtonian thinking dies hard.

The fact that the framework and functions of society haven't caught up in all this time is understandable when we realize that even the great Einstein had trouble wrapping his mind around the full implications of the quantum revolution. "God does not play dice with the universe" was his reversion to a kind of updated Newtonian science that had served him and all of science for centuries, and still does, though in a limited way. Until the end of his days, Einstein tried to find chinks in the armor of quantum theory.

Trying to get inside the mind of Einstein, I can only guess that he worried that we would throw out the baby (Newtonianism) with the bathwater. I suspect that he was not unaware of some of the worst aspects of outmoded Newtonian physics, which is still the basis of our thinking. The list is significant: fragmentation, disconnection, and rigid separation between peoples and states. Its deleterious effect on our lives, society, and science is profound and deeply tragic. We tend to isolate every problem, whether small or large. We want to get a quick fix, whether in war or research, rather than in concatenation with its context, causes, and effects.

In conventional medicine, for example, doctors will find a leaky heart valve or a spot of cancer and repair or cut out the offending part instead of treating the body as a whole. Just try asking your doctor about the role of nutrition in health. I'll bet that most will dismiss it. This is not surprising since medical schools, until recently, did not even include this subject in their curriculum except to recommend a "balanced diet," whatever that means. How else can one explain the dreadful diet served in hospitals?

Are You the Cause of Everything?

Quantum mechanics has led to arguments among philosophers, scientists, and pseudo-intellectuals. How can we change parts of the world just by observing them? Ridiculous! We "practical" people believe what our senses tell us. But in the subnuclear world, there are all kinds of magical events that our three-dimensional senses don't tell us about. Could we "tune in" to our deeper senses like the improvisatory jazz group? One is tempted to say that there is some kind mysterious, instantaneous communication between these players, just as in the two sister particles traveling far distances from each other and yet instantaneously "knowing" the position and momentum of the other. Carl Jung defined synchronicity as "the coincidence in time of two or more causally unrelated events which have the same meaning."

Much to the dismay of Einstein, synchronicities seemed to transcend the normal, testable, Newtonian laws of science, those earlier views based on a causally dominated world where nothing takes place that does not have a direct cause.

Causality has been likened to a falling row of dominoes, or a railway engine shunting a number of railcars. It hits the first car, forcing it to bang into the next, with the impulse traveling from car to car. Thus every event in the universe is causally linked to an event that comes before and to one that comes afterward. This makes good, everyday sense. Causal mechanisms describe a direct link or series of links between a linear sequence of causes and effects: ball one hits ball two, ball two moves ball three, etc. At the same time, it is important that we realize its limitations. This rather simplistic view does not apply, for example, in ecology, where the "causes" of change are infinitely more complex. Consequently, the search for deeper and more ultimate causes can lead reductionists into an infinite regress and can result in taking us further and further from our original goal. Infinite regression and expansion remove any possibility of mechanistic explanation. They do, however, provide a richly textured, highly complex albeit cloudy picture.

A team of scientific researchers would have a lifetime task trying to measure the minute differences in interpretation by the same orchestra of the same piece of music and connecting them to any number of causes, such as a variation of the conductor's beat, a change in the temperature of the hall, the smallest fluctuation in the oboe's A, any one player's minute changes of interpretation in the dynamics of a particular phrase and that player's effect on every other player and section, and so on. The researchers would have to pass on the research to their children.

Like the flapping of the butterfly's wings, everything causes everything else and each change in an event emerges from an infinite web of causal relationships. Even the earliest days of science were characterized by attempts to find pattern and order in the complex web of interconnections that exist in nature. Galileo made careful observations of balls rolling down a slope and along a flat table. In every case, the balls came to rest as a result of friction and air resistance. His quantum leap of insight was to realize that if these were totally eliminated, the balls would roll down forever. One of his major contributions was this realization that the natural motion of all objects in a straight line when they are not being acted on by another force is uniform.

Clearly, air resistance and friction cannot be eliminated in practice. They can, however, be eliminated in thought experiments, in the abstract world of mathematical physics. Galileo taught that in order to uncover the laws of patterns of nature, it is necessary to first abstract phenomena from the real world and consider laws in isolation from the contingencies of everyday life. It is only by performing thought experiments and abstracting out the contingencies of nature that underlying patterns can be inferred.

How do these patterns play themselves out in music? Let's take a symphony performance. It is the symphony conductor's task to weave together the incredibly complex web of orchestral dynamics, the context of which includes dynamics attributable to human beings and those emerging from the nature of different instruments, hall temperature and configuration, audiences, and the like. Then the conductor must cohere all the various elements in an ever moving and ongoing linear and vertical path to produce a result agreeable to the ear and to the composer's intentions. Add to this mishmash the conductor's interpretation, and you may have a righteous, Newtonian bore or a universally acclaimed, inspirational performance. Ideally, then, every performance of the same work must be fresh as though played for the first time!

Some decades ago a simple experiment designed at Canada's National Research Council tried to determine the acceleration due to gravity of a falling object, in this case a metal bar. The tube down which the metal would fall had to be evacuated to remove the effects of air resistance. Also, the drop pad had to be carefully shielded to remove all electric and magnetic forces on the bar. The whole apparatus had to be precisely temperature-controlled to eliminate changes in length due to expansion and contraction. Exact timing was accomplished by using an atomic clock. Even earth tremors, which would perturb the local gravitational force and would act to vibrate the bar, had to be monitored.

Chicken or Egg

With all the resources of this major research institute, the attempt to create ideal, insulated, and repeatable conditions in which to measure the acceleration to gravity resulted in each experiment differing slightly. If one expands this relatively simple experiment to try to determine the effects of the billions and billions of molecules in a gas or, in macro context, the effects of the moon on the gravity of the earth and vice versa or the effect of any heavenly body on another, we find an interminable series of interconnecting, interlocking, and interchanging developments where it is impossible to decide, as with the chicken and the egg, what the primary causation is. Some mathematicians virtually broke their computers in trying to depict, mathematically, the principle of cloud movements in order to accurately predict the weather.

They were not the first to try to reduce complex phenomena to simple principles. Leonardo da Vinci joined Copernicus, Kepler, and Galileo in the notion that universal principles were related to the external forces acting on the natural state of a body, causing it to move in a straight unperturbed line. It goes back still further to Aristotle's arguments that planets move in circles because such motion requires "minimum" amount of "effort." All these scientists were moving toward the universal principle, in which observed motion arises out of the most minimal effort.

Max Planck, discoverer of quantum energy, was so impressed by the power of these variation principles that he believed that they were universal, applying not only to the paths of moving bodies, but to all types of behavior within the universe. Since then, it has been shown how the theory of relativity and quantum theory follow the variation principles.

Beyond Newtonian mechanics, the flow of fluids and the behavior of electromagnetic fields arise out of the properties of the whole flow and the whole field, so that the variation principle is necessary to the understanding of a system.

A photographer once took rapid pictures of my baton in motion. Combining these in one photograph, one sees the baton in different positions from top to bottom, from upbeat to downbeat. If it were possible to take a sufficient number of photographs within the space of one downbeat, one would see that rather than a continuous flow of motion, the baton is actually moving in an infinite number of positions.

The world can be understood as being composed of interacting waves where motion emerges on the whole complex of dynamic movements of these waves rather than actual force acting on infinitesimal elements. This approach provides a framework for discussing field phenomena such as light and fluid and even suggests that material objects can be pictures of concentrations of wavelets emerging out of a whole wave field.

Can we accommodate the opposing mechanistic and organic accounts of nature as complementary to each other, to view causality and synchronicity not as mutually exclusive but as different perceptions that underly the same reality? If so, we may combine the subject and the object. This takes us back to the discussions among my fellow music students, as to whether to take an objective or a subjective approach to their practicing. The objective approach was to work on scales and arpeggios, and the subjective was to play and "feel" the compositions. From that, one gained the technique and muscle "intelligence" necessary to handle whatever technical difficulties the music might present. We learned that the best way is a combination of these.

This takes us into distant territory, namely the unity of consciousness in the universe, the bridge between mind and matter. This includes our thought processes as well, carrying us into treacherous territory, as in the current misuse of these phenomena, which poses significant dangers of thought manipulation by drugs, media, and politicians. It's a tricky time.

46

God, Appearance,
and Reality

Especially tricky is the search for the real as against
that which is partially real, totally imagined, or simply dreamed. The essence of research in physics and its
continual search for data is the search for reality—some
confirmation of a theory, something that can be proven
again and again in experiments held under all conditions. The same holds true for music. The music world
strives to probe the human need for resonance with the
right packages (be they song or symphony) of vibra-
tional frequencies to "sell" to themselves, record companies, and the public.

In art, the enemy of surrealists such as Magritte and
of true artists as well as musicians generally is naive realism—the dogged assumption that the human sensory
apparatus accurately records the one and only real

world, of which the human brain can make but one accurate model. To the naive realist, every view that does not fit the official model is dismissed. The naive realist sets himself up as the sole judge of what is real, settling on a single, unchanging representation of reality, and the gate slams shut behind him. He is doomed to live thereafter in the one universe of ideas or action to which he has pledged allegiance. His universe may be elegant, but it is a prison nonetheless.

The naive realist is only a hop, skip, and jump to a philosophical realist in terms of rigid pragmatism/conservatism. A homely example of both varieties of "realist" follows: Take an ordinary act in the course of a person's activities, like sitting, or walking to a definite place. These seem like simple acts that are done easily and that go relatively unnoticed. But in reality, these are highly complex, integral acts that are being formed at the level of infinitely small magnitudes, all interrelating and mobilizing at impossibly rapid speeds to achieve that "simple" act. As explained earlier, a performance on a musical instrument has the same perceptual appearance of simplicity.

Deceptive appearances abound in nature. The smooth surfaces of an ice cube, a planed and sanded wooden board, or a block of iron or marble are all deceiving. At the molecular level, the perfections are punctuated by imperfections, order by disorder. Looking inside the molecule, one finds a whirl of activity. Gone is the placidity on the surface.

Parmenides, of ancient Greece, saw the temporary as a mere illusory facade beyond which lay the eternal, the absolute. Though I reject absolutism in general, this concept is elegantly expressed by David Bohm:

> Each relationship has in it a certain content that is absolute, but this content must be defined to a closer and closer approximation, with the aid of broader concepts and theories that take into account more and more of the factors on which this relationship depends. The essential character of scientific research is, then, that it moves toward the absolute by studying the relative in its inexhaustible multiplicity and diversity.

The simple melodies of Henry Purcell or Christoph Gluck might be disjointed noise to a hearer whose partition of time is a thousand times subtler than ours. A person whose visual sense is far more acute than the norm would experience the "smooth" edge of a razor as being as jagged as an old, rusted saw. The cyber revolution can be reduced to ones and zeros.

In the decades since the inventions by Guglielmo Marconi and Alexander Graham Bell changed the world, the amount of data that can be transmitted has constantly increased. This has had a dramatic effect on the emotional quality of the message. The main reason that television can have more of an emotional impact than radio is that it communicates more data. An AM radio signal, for example, requires only ten kilohertz (ten thousand cycles per second) of bandwidth, while a high-fidelity FM transmission takes two hundred kilohertz and a color TV signal takes about six megahertz (six million cycles); color television therefore conveys six hundred times more data per second than AM radio. Whether the data is better is another matter!

The progression from drumbeats as communication to live television broadcasts can be summed up as the relaying of more and more data, with less distortion, in less time. The prospect of real-time trans-global interaction makes the world seem very small. Every single person is a potential neighbor. The recognition of this is a challenge to the music industry and to politics. The field of physics leans heavily on the data gleaned from observational and computational data, expressed in equations and then introduced into computers. It is an exciting, involving, open-ended world that has teachers talking of a new era in education while also lamenting that they are forced to teach scripted and restricted lessons and that their students spend too much time playing video games.

Absolutists are far more rare among musicians and music lovers than among religious fundamentalists. Music lovers usually have a particular genre that they favor but are generally open to experimenting in other genres. Fundamentalists, on the other hand, are absolutely certain they are right and all others are either wrong, misled, or, most likely, sinners. The Bible is *it,* word for word. Can't be wrong. "There it is in black and white!"

47

Absolute Truth: Design Versus Evolution

The centuries-old discussions of design versus evolution have been given new life in the past few years. Newton's "laws of nature" led to assertions that the perfect harmony of nature's laws pointed to the existence of a legislating deity. Eighteenth-century philosopher Immanuel Kant was an enthusiastic proponent for a scientific description of the world, based largely on Newton's law of motion and gravitation. Kant proposed a theory for the origin of the solar system, and held the view that there was a real world out there, only then to be described by our minds. Though Kant's arguments undermined the belief in the existence of God, he was sympathetic to the design arguments, trying to convince his readers that we cannot use the evidence of our senses or indeed our thoughts to draw reliable conclusions about the ultimate nature of "true reality."

Kant's perception of space was influenced by the belief in the absolute character of Euclidean space. This is the geometry of lines on flat surfaces that we all learned at school. It is characterized by the fact that the sum of the three interior angles of a triangle are always equal to 180 degrees. Euclidean geometry was considered reality until scientists and astronomers discovered that there could be logically consistent geometries that differ from Euclid's conception. These non-Euclidean geometries describe the properties of lines and curves on a surface that is not flat, and of triangles that are constructed from the shortest lines between three points and do not have interior angles that add up to 180 degrees. Einstein confirmed observations that the underlying geometry of the universe is non-Euclidean and that curved surfaces look flat only when viewed over sufficiently small regions. For example, the earth's surface is curved, but seems flat when we look at short distances. However, when we observe accurately over large distances, the curvature of the horizon becomes evident.

This discovery eroded the confidence of theologians and philosophers in the concept of absolute truth. It is interesting to note that as early as the fifteenth century, artist Jan van Eyck painted the Arnolfini wedding portrait, where the couple and their dog are perfectly reflected in a convex mirror hanging on a wall behind them. The perspective is complicated by the use of more than one vanishing point. Artists such as van Eyck and Leonardo have long predicted the future.

There are movements in theology, as in physics, that teach that the universe is running down like a great Victorian engine. "Process" theologians developed the concept of an evolving God who does not know what the future holds. We find a sharp distinction between the theologians who regard the presence of time and the flow of events as being of vital theological importance and some modern cosmologists who see the future as already laid out and determined—the ultimate "running down" of our universal clock. Can God wind it up again, or is He also an absolutist?!

Lonely in Our Universe

The concept that we are alone in the universe reduces our sense of the vast reaches of space, time, and space-time, as well as the infinitesimally small microworld. And it has led us into all kinds of superstitious beliefs.

Mystical, symbolic, and religious thinking seems to characterize human thinking everywhere and at all times. Religious belief encompasses countless deities accompanied by a multitude of different rituals and beliefs. The relief and comfort that emotional fervor and blind belief has given to so many would be enviable if not for the fact that it too often robs the world of rationality, wisdom, and constructive action.

I have a relative who is totally rational about his work and daily life and yet is a passionate proponent of literal truth in the Bible, wherein every word tells how things are. When one tries to counter with what seem to be incontestable arguments and with plain common sense, it is all to no avail. Adam, Eve, and the apple win out, science loses.

48

A Crisis in Physics

The twentieth-century crisis in physics, never fully resolved, is sometimes held to consist of the contradictions between macroscopic or relativity physics on the one hand and quantum or atomic physics on the other.

The followers of Einstein and Max Planck in the relativity corner and Heisenberg, Schrödinger, Dirac, and Eddington in the quantum corner reached an uneasy truce when string theory came on board. Those scientists espousing one or the other conceptual approach realized, even before string theory, that they needed each other. If a grand unified theory was to be found, the macro and micro worlds would have to be reconciled, and included with them would have to be that ever troubling force, gravity.

On a related front, Einstein and Planck's causality approach has been fighting what sometimes seems to be a losing struggle with those in the theological and political domains, with a sprinkling of writer supporters, who interpret quantum mechanics and chaos theory in a way that lends legitimacy to their denial of causes for any event. Instead, they leave it all up to God. But the theory is too strong to be snuffed out or distorted, especially in a time of social and political crisis.

This particular contradiction can be traced back to a time when seventeenth-century Dutch physicist/astronomer/mathematician Christian Huygens proposed that light traveled through space as a wave transmitted by an invisible substance called the "luminiferous ether," while Newton put forth the corpuscular theory of light, asserting that light is made up of tiny particles that shoot through space like billiard balls, in straight, single-file rays. Newton formed a consistent scheme of the universe, built up on an atomistic basis. All particles behaved according to a single law of motion.

God Intervenes

Newton's system was of such a character that an "initial push-off" and a fabrication of the atoms out of nothing was necessary. These initial acts were creative acts of God, who thus appears as force and substance alienated from Himself. All things, once created, are subject to the laws of the conservation of matter and energy, and after being given its initial push-off and creation, the atomistic universe is self-running. God can take a vacation.

Newton, however, did not quite regard it in this light, though his conception of substance was such that, whether he realized it or not, the maintenance of these laws would require the continual intervention of God. Thus, such a universe does not exclude the possibility of divine interference with its own laws.

Aristotle and Contradictions

In the medieval and Aristotelian schemes, motion requires the constant expenditure of force. Hence the universe needs the continual inflow of divinity to keep it going. Newton's atomistic scheme gives a basis for deleting God as a causal influence once the universe is taken into consideration. The laws of God then become qualities of matter. Another contradiction.

As compared with Aristotle's theories, Newton's laws of motion desacralized physics, culminating in Laplace's divine calculator, which, if given the speed and location of every particle in the universe, could predict the whole future course of events. Nature becomes a machine. Newtonian physics excludes God from nature but not from reality, because, as a result of its particular conception of matter, it views nature as being only a part of reality.

The "proof" of the corpuscular theory of light and Laplace's divine calculator were shattered when it was proved that light consisted not of corpuscles but of waves. Of course, waves have to be waves of something, as in water waves, but waves of light were waves of nothing until the emergence of the ether theory. This theory claimed that what looks to us as nothing, or as air, is made up of a substance that they labeled "ether."

Our high school physics takes the story from there, except that the denouement reached today gives little satisfaction toward a permanent solution of the basic contradiction. The crisis in physics is not just a scientific, technical, or philosophical crisis. As we have seen, science cannot be separated from the crisis in economics and in human affairs in general. Perhaps delving into the past may give some clues as to how the crisis can be resolved.

Confucius Screams at Lao-tzu

In China in the sixth century B.C.E., there was a similar dichotomy between Confucius and his contemporary, philosopher Lao-tzu. It has been said that when Confucius visited Lao-tzu, he emerged from the meeting screaming, "This man is dangerous, and you must have nothing to do with him!" A stark difference divided the two famous men. In essence, Confucius taught: If you are a slave, be a good slave and be satisfied with your place in society. The world will work fine if everybody stays where he is.

Lao-tzu taught the direct opposite, though his philosophy finds interpreters at odds with one another about what they see as his true message. According to Dr. Ralph Alan Dale, author of *Finding Our Dreams in Lao Tzu's Tao Te Ching* (now published by Barnes & Noble as *Lao Tzu*), Lao-tzu was a rebel and a visionary. The following excerpts are Dale's interpretation, applicable for today:

> Those on the path of the Great Integrity
> Never use military force to conquer others
> Every aggressive act harvests its own counter-terrorism.
> Wherever the military marches,
> The killing fields lay waste to the land,
> Yielding years of famine and misery.
> Aggression leaches our strength and humanity,
> Subverting the Great Integrity,
> And inviting disaster.
> After the battle, the soldiers who have slain others,
> Move to the yin side and mourn.
> Even while the commanders, now on the yang side,
> Are celebrating victory, which is always a funeral.

Lao-tzu is said to have counseled his followers to follow nature rather than current social-structure hierarchies.

49

Social Change
According to Einstein

"The economic anarchy of capitalist society as it exists today is, in my opinion, the real source of the evil. . . . Private capital tends to be concentrated in few hands . . . [resulting in] an oligarchy of private capital, the enormous power of which cannot be effectively checked even by a democratically organized political society. This is true since the members of legislative bodies are selected by political parties, largely financed or otherwise influenced by private capitalists. . . . The consequence is that the representatives of the people do not in fact sufficiently protect the interest of the underprivileged sections of the population. Moreover . . . private capitalists inevitably control, directly or indirectly, the main sources of information (press, radio, education). It is thus extremely difficult and indeed in most cases quite impossible for the individual citizen to come to objective conclusions and to make intelligent use of his political rights.

The crippling of individuals I consider the worst evil of capitalism. Our whole educational system suffers from this evil. An exaggerated competitive attitude is inculcated into the student, who is trained to worship acquisitive success as preparation for his future career." —EINSTEIN, 1949

As a musical person, Einstein was especially sensitive to prescience and precognition. His sentiments echo today in a period when many countries are going through social and economic turbulence with related religious and racial strife. Surely Einstein would have agreed that time has run perilously short. Capitalism's market economy depends on competition, profits, and the satisfaction of greed. Individual profiteers of the system cannot be wholly blamed, because even the few well-meaning captains of industry are locked into a paradigm wherein "if we don't do it, our competitors will." Inevitably the values and character of human civilization are besmirched. They become willing participants in their (and our) own demise.

As we work within our market system, our decisions are made for us in the system's effort to perpetuate itself. The rush toward cheaper labor, for example, whether in China, Mexico, or India, becomes a necessity in order to meet "the bottom line." American jobs are lost. Corporate leaders of industry and government collect skyrocketing profits and shrug their shoulders. Developing nations, happy to garner a few jobs, are despoiled of their natural (and human) resources; some eagerly advertise their willingness to be used in this manner. A climate of national nervousness pervades the rich industrial societies. Our time to act continues to dwindle, but which way do we go?

Economists and politicians have few answers. The inevitable toxicity of this global system affects a starving child in Bangladesh as much as it does the child of a corporate capitalist or, for that matter, a media pundit or self-important economist.

Can music and physics hint at a road map to take society out of chaos? Characteristically, both disciplines tend toward creating order and unity out of chaos. The physicist sorts out a seemingly chaotic universe, and the musician produces order out of disorder. The composer, the painter, or the sculptor chooses from a universe of materials, eliminating that (from marble or sounds) which does not fit with their conception. Social chaos is no mere theory; its effects on people, from the poorest nation to the very richest, present us with a future bleak and desolate—unless we first examine its roots and then take appropriate steps. Let's look at the roots.

Unreality and Fragmentation

In recent centuries we have been living in an unreality that has resulted in crisis after crisis. Today's is the culmination of all previous crises. It pervades every aspect of society, music, science, politics, and the earth itself. We can choose between two conflicting ways of looking at the world:

1. By way of the old physics and its fragmentation, quantification, and competition, or
2. By the new physics of relativity, quantum mechanics, cosmology, and string theories. Though this "new" physics is already a century old, we have yet to catch up to it in our thinking and practice. Instead, we are floundering in a paradigm of obsolescence and violence.

Violence of one kind or another is no stranger to me. I have seen the world through world wars and many "smaller" wars; I have watched as scientists harnessed technology, reaching for the stars while plunging whole cities into conflagration and endangering the entire planet. I have experienced the exhilaration and the letdown of utopian ideals and depressing failures despite the Promethean efforts of many to bring the world to a state of peace, prosperity, and enlightenment. The same technology that bequeathed us its benefits also promotes the destruction of self and species.

My interest in physics arose out of my sense of frustration at seeing the sheer idiocy of a world being run by way of long-dead paradigms of political structure and brutal social behavior, both of which arise out of obsolete societal and human relationships. It was antithetical to the logic I saw in music. Physics substantiated notions I had heretofore only intuited, ideas about holism, the links between disciplines, and the implications for relations between peoples. It confirmed my convictions about the connections between misdirected science and social ills. It seemed to me that racism, for example, was just another form of malignant fragmentation, the separation of peoples into superior/inferior, controller/ controlled, master/slave, and exploiter/exploited. Religious fragmentation was responsible for bloodbath after bloodbath.

Cut and Burn

Fragmentation was everywhere I looked. Medical patients are treated locally, with little or no connection between the part and the total bodily system. If a part offends, we cut or burn it out (as in invasive operation, chemotherapy, and radiation), with little consideration for side effects or, better, the science of prevention. The good news is that the message is getting out. More hospitals have added "complementary medicine" departments to meet the demand of educated patients. Some hospital cardiology clinics offer the Dr. Dean Ornish (Clinical Professor of Medicine Research Center, University of California) "lifestyle change" program, which includes dietary modification, exercise, group support and sharing, and yoga and meditation, as well as the avoidance of addictive poisons such as tobacco, processed sugars, and other drugs. The "lifestyle" value has been proven in rigorous randomized clinical trials. I'm a case in point, having successfully followed a similar regime after being offered dire "six months to live" diagnoses three times by "the best doctors"—in my teens, in my thirties and as recently as fourteen years ago.

In science, fragmented innovations tumbled into use one after the other, often with little attention to the long-term effects on humans and our lonesome, precious planet. The benefits of science were fast being destroyed by this schizophrenia.

The twentieth century saw music become still another manifestation of fragmentation, with new genres and approaches continually spinning out: In addition to the old "classical" music based on the harmonic series, there was the 12-tone scale, "modern music" (singularly audience free), aleatoric (music of chance), highly structured (electronic or mathematical) music. Efforts to create something new and different in so-called serious music were matched in electronic music, new age, rap, rock, minimalism, jazz, folk, gospel, ethnic, and stews of all these. Fresh new music in all these groupings did manage to break through, in the music of the Beatles, Sting, later rock and pop musicians (some of the most direct social messages came via pop-song lyrics), and some "classical" composers such as John Adams, Ellen Taaffe Zwillich, John Corigliano, and Philip Glass.

Music's Social Context

My seventy-odd years in music have honed my realization of music's mystery and awesome power. Inescapably, I became involved in pursuits that explored music as a vehicle for social change. Music and social change animated my life as I discovered music's power to reach truth in ourselves, a power it has in all cultures.

Superficially, it would seem to be unthreatening to the power structure because of its abstract nature. Yet both music and science have always made the power structure nervous. During World War II, General Grove, head of the atom-bomb Project, was uneasy about the egghead scientists he had to engage (including the greatest scientists of the century!) in the making of the bomb. Like science, music *happens,* irresistibly, for good or evil. Italian opera, essentially opera of the people, is embraced by the elite. Verdi crossed class lines; his *Requiem* put heart and soul back into the people, and his opera debuts often afforded a gathering for nationalist revolutionaries.

Religious music (for me, *all* music is religious) has served as a compensation for the desperation that people feel in their lives. The dirt-poor Catholic masses in South America and the spontaneous religiosity in a Southern African-American church are enchained to the Church by the magnificent music. The music revives the celebrants' heart and soul squeezed out in their everyday lives. Even a staid concert-hall performance can be a religious experience.

For years I didn't attend performances of the New York Philharmonic, despite the fact that the internationally famous conductor was extremely capable. Capable wasn't enough for my colleagues in the orchestra; there had to be genuine communication from within, a revelation about some portion of the human condition. Call it spirituality or emotion. When the New York City Ballet performed in the famed Marlinsky Theater in St. Petersburg, Russia, during the week of August 4, 2003, one longtime fan's response was: "I didn't feel any emotion. They did their job, but it didn't stir up any enthusiasm. It was cold."

Naturally, these technological times have their reflection in the arts. A premium is set on "professionalism," virtuosity, and, in the case of many conductors and performers, speed. Professionalism is related in history to the growth of large-scale commercialization that began roughly with the industrial revolution. Some music, reflecting these fast-paced times, can be fast, but why apply that pace to all the wondrous repertoire of previous centuries? Can it be fear of the messages inherent in the deeply human music of Beethoven and other greats of the eighteenth and nineteenth centuries?

50

Music's Dangers

The fear of music has a long tradition. In the fourth century, St. Augustine confessed, "I have become a problem unto myself," as a result of his "sensuous" pleasure from the melodies he heard in church. John of Salisbury worried that the music in the Notre Dame Cathedral could "more easily occasion titillation between the legs than a sense of devotion." Policemen of music in England and Switzerland burned books containing "popish ditties," and the Orthodox Patriarch of Moscow ordered bonfires of musical instruments. His church complained about "dancing, and cymbals and flutes, and shameful songs from the lips of painted girls."

Political control through music is nothing new. The notion of controlling music extends back to early civilizations, goes through the Enlightenment, and continues to the present day. Socrates, ostensibly devising a perfect state, argued that certain Greek musical modes should be forbidden. The Lydian mode must be dispensed with, Socrates noted in Plato's *Republic,* because it would encourage "wailing and lamentations." Other modes, such as the Ionian, were considered too slack for use in training a group of warriors. Educational music must be carefully controlled because rhythm and harmony "insinuate themselves into the inmost part of the soul." Later, the tritone, or *diabolus musica,* was to be avoided because of the sound it would have introduced into a devotional setting. Something as simple as an interval was feared to invoke a world beyond the ordered system of rational composition, suggesting something threatening and unpredictable.

The New York Times Arts and Leisure section of December 9, 2001, featured a full front-page article headed "Music's Dangers and the Case for Control." Richard Taruskin expounded at length on the history of "the enemies of music, like the Taliban, [who] fear it has power over the soul," though he waffles with: "Censorship is always deplorable, but the exercise of forbearance can be noble. Not to be able to distinguish the noble from the deplorable is morally obtuse."

I wonder who Taruskin would choose to be the arbiter between noble and deplorable music? Would the appointed Committee include the Ayatollah Khomeini (who banned music from Iranian radio and television in 1979 "because its effects were like those of opium, stupefying persons listening to it and making their brains inactive and frivolous")? Or Senator Tom DeLay, Congressman Pete Hoekstra, or former Senator Jesse Helms?

Taruskin recalls his early days as a college music teacher when "a young Hasidic man in fringes and gaberdines approached him asking to be warned in advance" when a record would be played that contained the sound of a woman's voice, so that he could slip into the hall and avoid it.

Citing Socrates's banning of most of the modes or scales, Tolstoy's comparing music's effects to those of hypnosis, and Plato's "mingled awe and suspicion of music's uncanny power over our minds and bodies,"

Taruskin posits "music's affinity with our grosser animal nature" as the reason for "musical modernists to put so much squeamish distance between their cerebral art and viscerally engaging popular culture."

Then, as is fashionable in some intellectual circles, Taruskin's "moral equivalency" couples the Soviets with the Nazis. He brings up the forties, when Soviet cultural minister Andrey Zhdanov castigated Shostakovich and, to a lesser degree, Prokofiev, for their musical "formalism." In the case of Shostakovitch, at least, the world can be grateful for that period of official criticism, for out of the little-performed and much-criticized symphonies 2, 3, and 4, and the opera *Lady Macbeth of Mdensk,* we are bequeathed the great Symphonies 5 through 15, his fifteen string quartets, and other splendid works. So there may be some value after all in knowledgeable criticism of the arts if not connected to artistic castration.

The debate about Shostakovich has been ignited again in reissued publications by Taruskin and Volkov, both of whom gain little currency with this writer/musician, who did a thorough study of Shostakovich years ago, concluding that the composer *was* inspired by the Soviet dream—though with eyes opened by the horrific realities at the end of his life. He was a Russian patriot but far more, he was a passionate socialist, dreaming of a better world, like so many in those days. In the unlikely case made by Volkov and others that he was only dissembling to fool the government, then the nuances of his writings matched his skill in musical satire. In the forties, fifties, and sixties, I listened avidly to his music and read everything I could find about him, his music, and his own writings on political, musical, and social issues, and I largely shared his views, for he was, yes, a committed people's socialist. I recorded his Symphonies 5 and 15 with the London Philharmonic for Charisma, a British *rock* record company! These are now long out of print, for all I know. In the Fifteenth he quotes from Wagner's *Ring,* although he had little use for the nineteenth-century composer of "bourgeois-fascistic tendencies." In his Seventh he depicts with bass and percussion the marching Nazi hordes in World War II with shattering effect.

The taboo on performance of Wagner's music in Israel was broken by the great conductor/pianist Daniel Barenboim, who was roundly crit-icized by American and Israeli Jews. Mr. Barenboim, himself a Jew, felt that even though Wagner the man was notoriously anti-Semitic, his mu-sic clearly merited performance. Music was held hostage while the real creators of the Holocaust, the German Nazis of WWII, were quickly ac-cepted as friends and partners of the U.S. and even many in Israel. Pales-tinians continue to be subjected to brutal treatment as though *they* had committed the Holocaust atrocity!

Music's power to light the way to truth also has the power to make people uncomfortable. A lump in the craw of power structures is that most artists in the entertainment world are liberal, hence dangerous in the eyes of those who want only to protect their positions of privilege. Artists who become celebrity heroes and heroines may use their art and status to educate and move the public to vote out regressive government leaders, a frightening prospect for entrenched officeholders. Even though government/corporate power may not be able to totally stamp out or corrupt freedom of the arts, television and the print media are doing the job for them by a general dumbing down of their viewers and readers.

51

Arts and Science Power

The power of the arts is revealed by its enemies. Jane Alexander, Chairman of the National Endowment for the Arts from 1993 to 1997, tells of her experiences in her book *Command Performance: An Actress in the Theatre of Politics.* Ms. Alexander, no hotheaded liberal but a highly regarded, award-winning actress, is a proper New England lady of considerable aristocratic dignity. She relates her struggles to convince conservative Republican and Democratic senators to at least maintain a meager budget for the arts institution. Powerful senators and congressmen such as Strom Thurmond, Orrin Hatch, Charles Stenholm, and Jesse Helms pasted a label on the agency as a "mist of homosexuals." One would then conclude that some of the most established arts organizations in the country, such as the Metropolitan Opera and most ballet companies, are hotbeds of deviants.

There was discussion in Congress of eliminating the entire agency or radically reducing its budget, already so minuscule it could easily hide in the corner pocket of almost any other agency, never to be found again. The NEA barely survived, with a budget much smaller than the national arts budget of any industrialized country and even that of many poor countries. The level of debate in the Congress was so debased that it became the laughingstock of countries such as England, France, Russia, and Germany, not to mention thoughtful people in the United States.

In all the years of her tenure as president of the NEA, Ms. Alexander was never able to have a face-to-face meeting with North Carolina senator Jesse Helms, who held a key post that allocated government funds for the arts. Alexander relates an encounter her much-acclaimed playwright friend Wendy Wasserstein had with Senator Helms. "Senator Helms, I would not have had the opportunities to write and produce *The Heidi Chronicles,* which won a Pulitzer Prize, and my other plays if it hadn't been for my grant from the NEA. There are many talented playwrights whose work could be furthered with this funding."

"Well," replied the senator, "look at all the great plays by Arthur Murray. He never had an NEA grant."

I'm sure the reader knows that Arthur Murray was a promoter of ballroom dancing! He had nothing at all to do with writing plays.

In her frustration, the normally polite Alexander reacts uncharacteristically: "Tom DeLay's brain could fit in a cup as far as I could tell. . . . I felt contempt for many of the Members [of Congress] and despair over the level to which our government had sunk." Pete Hoekstra, Michigan Republican, evokes this from Alexander: "His tactics were scurrilous, his aide was rude and underhanded."

Defeated, she adds: "The system is so corrupt, it may never be fixed."

Tainted Music?

A generation or two of young punk, rock, rap, and popular musicians, many classically trained, encompass a wide range of expression, from utopian ideals to revolutionary thought. Their messages contain elements of social protest and, most recently, a perverse kind of racism and sexism, which in the rock world's private language is itself a kind of protest. The award-winning pop star Eminem was rarely understood by adults until his huge following, record sales, and hit film—*8 Mile*—led to the business pages embracing him as one of their own. Tom Morello's band, Rage Against the Machine, overtly political, anticorporate, and "totally subversive revolutionary propaganda," sold 14 million records and was released on Sony! Business is business.

The German pop star Nina Hagen boasts of having once masturbated on an Austrian television show. She has had triumphant tours in many countries except the U.S. Her growling, screaming, punk-rock mixes operatic excerpts with clichés about peace ("the universe is as big as love") and animal rights.

British folk-rocker Billy Bragg scored considerable airtime with a song about the global justice movement. The first single from Bragg's *No Power Without Accountability* is destined to become an enduring anthem for anticorporate organizers. Bragg is not just a popular musician; he has many ideas about the negative impact of globalization and the duty of a political songwriter. He was criticized during the Margaret Thatcher leadership in England from 1979 on. Bragg didn't vote that time and says that perhaps that was the arrogance of youth. At the height of punk, he called himself an anarchist, though he admits that it was hardly a developed idea. But as Thatcher cut chunks from free health care, free education, decent affordable housing, and the like, he became more politically involved, especially during the 1984 miners' strike. So his group started doing gigs in the coalfields, and he began to define himself by something other than the "blowin' in the wind" sort of politics. He believes that "if you want to change the world, you should be organizing fast food joints," and "a positive way of changing the world, though slower, which won't get you on CNN, was the sort of campaign that I worked with in the U.S.A.—Justice for Janitors."

He calls Woody Guthrie "the father of my tradition—the political singer/songwriter tradition. Where the text is about unions, Woody's songs get on the charts. . . . The job of the singer/songwriter is to reflect the world around him and obviously the global justice movement.

"I went to a Methodist church where activists were speaking about how they were going to organize demonstrations around the World Economic Forum. They asked me to sing the 'Internationale.' That really touched me, because we do have a strong tradition on the left and . . . we have an opportunity to create a leftist idea outside the shadow of totalitarianism."

In late December 2003, for the last part of a three-week, thirteen-city tour, two Federal Communications Commission members, Michael J. Copps and Jonathan S. Adelstein, joined with touring musicians Bragg, Steve Earle, Tom Morello, Boots Riley, and other prominent rock stars in a "Tell Us the Truth" tour, aimed at educating the nation on the perils of media consolidation. Media deregulation has been debated on college campuses and in major print media ever since the Recording Industry Association of America convinced Congress to insert a small clause into the copyright law that would have given record companies song rights in perpetuity instead of having them revert back to the artists. Within a year the Recording Artists' Coalition succeeded in having the clause repealed. Next, musicians including Sheryl Crowe and Don Henley testified before Congress, hundreds signed a petition to the FCC, and the Future of Music Coalition commissioned a report showing how media consolidation has hurt diversity and restricted songs, and even censored songs such as John Lennon's "Imagine." Two radio conglomerates, Cox Radio and Cumulus, banned the Dixie Chicks after a member of the group, Natalie Maines, said she was ashamed to be from the same state as President Bush. Other musicians have said that they are afraid to speak out, sign petitions, or write controversial music, imposing a self-censorship damaging to the creative arts.

Are these sensitive artists a threat to society? On the whole, much of their music and language are no more than typical youth rebellion clothed in preachy but nonthreatening anarchy. Most are good-natured, shy, and diffident, like George Harrison. Some of the greatest rock artists—such as John Lennon, Bob Dylan, and Neil Young—are vaguely utopian. This doesn't stop the wheels of power from fearing them and trying to steamroll over those artists and their music.

Lennon, peacenik personified, was hounded by New York City and U.S. officials aiming to deport him. At the request of Yoko and John, I joined in preventing this outrage by recruiting the then deputy mayor of New York, who spoke up for John, joining Lennon's thousands of loyal supporters. Outcries across the land were successful in keeping him here.

Protest music was also written by skillful 12-tone composers, such as Luigi Nono, who composed *Tre epitaffi per Federico Garcia Lorca,* a monument to the Spanish Civil War, and Luigi Dallapiccola, who wrote *Canti di Prigionia* and *Canti di Liberazione,* both celebrating the end of Hitler's Nazi era.

A few years ago, the nude sculpture of the great tennis star Arthur Ashe was removed from Flushing Meadow. Eric Fischl's sculpture commemorating September 11, *Tumbling Woman,* was removed from Rockefeller Center. It reflected his personal horror and trauma. Both apparently offended the arts police.

Perhaps the conservatives are on to something. Although they fear that music (defined very broadly by this writer to include dance, the visual arts, and much more) might spill over into social action, the "culture industry" (Theodor Adorno) as a whole finds that dissent is quite marketable and profitable. Paradoxically, too, while artists are heroes to millions of people, especially the young, most artists tend to feel impotent against their employers—the record companies—and the forces of government. Both left and right are becoming cognizant of the power of the arts to sway opinions.

Record companies are sometimes unable or unwilling to deal with artistry and creativity in music. A rock band like Wilco, for instance, might create an album like *Yankee Hotel Foxtrot,* which was considered by many in the mainstream music press to be a landmark album in 2002. *Rolling Stone* called it the year's first truly great album. And yet Reprise Records, the band's label, found the album troubling; they did not understand it, and did not want to understand it or give it the time necessary to enjoy it for the work of art it is, and were unable to discern a target market for it. So they refused to release the album, which was eventually released on Nonesuch, a label known more for its classical output.

The funny thing about this is that Reprise paid for the cost of recording the album, and when they decided not to release it, they sold it back to Wilco at a bargain rate. Wilco then sold the album to Nonesuch. Because both labels are owned by the same corporate megalith, Time Warner, in effect, Time Warner bought the record twice, for three times the money it would have cost them otherwise. The artist, for once, was victorious over the business. A rare exception to the degenerate record business is Artemis Records, whose president, Danny Goldberg, is not only an extremely gifted judge of good music but is fair in his treatment of artists and is a politically active social visionary.

Music Marches On

The revelations of science have exposed the glaring dichotomy between the amazing discoveries in physics and the low levels of thinking displayed by government and other leaders. Einstein and other great minds—musicians, artists, philosophers, and scientists among them, have long deplored this situation. Many scientific revolutionaries come to mind. Galileo was arrested and spent his last years under house arrest for defending the Copernican system and for his belief that if people weren't allowed to question and interpret the meaning of scripture for themselves, "why would God have given them a mind?"

We don't have to go far back to see examples of artists and thinkers who suffered for advocating liberty and human rights. I'm proud to include myself in the long list of educators, entertainers, and artists who were blacklisted during the McCarthy period in the early fifties. This period mimicked earlier witch hunts. Victims included many prominent actors, producers, professors, musicians, and teachers. Playwright Arthur Miller was forced to appear before the Congressional Committee. President Eisenhower, Republican conservative and war hero, was charged with communist sympathies and being a "dupe" of the communists. Some wonder whether that era is being repeated today. The songs and visual arts are mirroring these social developments and while some scientists busily upgrade the biotechnology of war, others, especially those in basic research, find their subsidies dwindling.

As music and art are cut from school curricula, one wonders about the loss of young musical and scientific talent in our country. In my case, to compensate for the limitations of my early home environment, I found exciting ideas and people in voluminous reading. Since there were no books at home, I discovered the treasures of the local library. It became my refuge from a noisy world. Thank God for Andrew Carnegie. Music soon brought me out of my private, introspective world as I gained recognition by high school teachers in the band and orchestra, and was selected to perform for the student body at high school graduation.

A science teacher also contributed toward bringing me out of my lonely world. This high school chemistry teacher invited me to join two other students in an extracurricular, private class he gave on college-level organic chemistry. On graduating I was thinking of a career in science but there was no money for college. Our family "elder" advised me that there were quotas for Jews seeking to study medicine or science. Furthermore, I had no inkling as to how one applies for scholarships or funding, nor did I know about colleges. In those days, there were no career counselors in the schools. More relevant, because I grew up during the Great Depression, there was no money for me to go to college. My unschooled immigrant parents could give no direction or financial help.

An elder sister, Gertrude, came to the rescue. Then living in Philadelphia, she was dating a member of the Philadelphia Orchestra, who managed to get me an audition at the prestigious Curtis Institute of Music, where I won a full scholarship. After completing the full curriculum in less than three years, World War II beckoned.

52

Eleanor Roosevelt and
Our Nation's Capital

I was very anxious to defeat Hitler and the Nazis, single-handedly if necessary. After being turned down by several U.S. services, I managed to enlist in the U.S. Army Air Forces Band in Bolling Field, just outside Washington, D.C. This elite band was drawn from musicians in Toscanini's NBC Symphony, the Boston Symphony, the New York Philharmonic, and famous jazz big bands. We played for the troops, the generals, the president, and the public. We were also stationed in England's Uxbridge military headquarters outside London and later, in France. I won an essay contest on the theme "Why We Are Fighting the War" and thereby met the officer in charge, Mark Stone, and he and his family, including Izzy Stone of the *I. F. Stone Weekly* and Judy of the *San Francisco Chronicle,* became lifelong friends of mine.

Washington had several servicemen's canteens, but none for African-American soldiers and WACs (Women's Army Corps), despite the fact that African-Americans were sent to the most dangerous zones. A buddy, Sergeant Micky Salkind, and I decided that something should be done about the canteen situation. Before the war, Micky had worked in the government. He introduced me to his former CIO union representatives and succeeded in getting me a slot to speak at their next meeting. After my pitch, asking their support for a canteen that would not discriminate, they voted to give support, and the CIO Canteen was born. This was years before the CIO (Congress of Industrial Organizations) merged with the other major union, the American Federation of Labor (AFL).

We found a rundown loft and began fixing it up. Some Southern congressmen heard about it, found a spurious excuse, and had the fire department kick us out. We put out a cry of despair to Eleanor Roosevelt, and she immediately came to the rescue with a five-story building on Connecticut Avenue in a posh, central area of Washington. We were in business. We recruited quite a few civilians and service people for our cause and planned a full spectrum of activities. On one floor we served food and drinks, another was devoted to dancing, and another was where people played games (cards, chess, checkers). We even developed a training program for hosts and hostesses and put out a newspaper, *The CIO Canteen.*

Mrs. Roosevelt would come and serve on the food line. She was a gracious hostess and worked humbly, just like any of us. What a beautiful soul she was! The same congressmen who had previously managed to evict us from our former quarters made speeches (recorded in the Congressional Record) stating: "Our fair flower of southern womanhood is being besmirched." Throughout the life of the canteen, no one was "besmirched" (I love that word). There was never an ugly incident, and it was said that we "broke the back of Jim Crow in our nation's capital."

Youth Arts Forum:
The Future of Art

Mustered *out* of the U.S. Army Air Forces Band in September 1945, I decided to write a letter to *The New York Times*. It was published in the Sunday, November 18, 1945, edition, under the headline ARMY BANDS-MEN EYE FUTURE OF ART.

Here are excerpts:

> The first hand experiences of service musicians returning from over-seas are bringing into sharp relief the question of America's cultural future.
>
> After the first haze of our homecoming wore away we began to make some comparisons with European countries and came to some interesting conclusions. The America we had come to idealize did not disappoint us in its abundance, efficiency and great wealth. The automobiles, cigarette lighters, ice cream sodas and oranges were plentiful, ingenious and gratifying. Even the illusion of American inferiority in music and the arts was completely shattered. Not only are our orchestras and performers technically, and in many cases artistically, superior, but even our creative talent, so oft maligned, is endowed with a vigor and imagination that needs no apologia.
>
> Here, however, we found an almost incredible contradiction between our abundant talent and the use made of it. Noticeably lacking was, and is, the recognition and exploitation of art and music as an integral part of ordinary life, and not just a luxury. We were reminded of the prominent European musician who charged that we Americans think with our stomachs.
>
> In Europe, even in the midst of deprivation and destruction, the governments strove to meet not only the physical wants of their people but their cultural needs as well. England in 1944 in the midst of the buzz bombs raining on their heads (and ours!), allocated £175,000 for music. France was even more generous, as they spent

millions of francs annually through their Ministry of Fine Arts, for theatre, opera, conservatories and symphony orchestras. The liberation of Poland gave rise to a conference of the Government and its foremost artists for sponsored tours and cultural exchanges.

Why, then, in the United States do we permit an antiquated laissez-faire attitude that stifles our artists and puts our musical future in jeopardy? Is it because the American people are unmusical? The experience of all service and USO entertainers belies this emphatically. GI audiences crowded hangar and hall to listen with the same eager attention to Bach and Tchaikovsky as to boogie-woogie.

Your recent suggestion that a Federal bureau of fine arts, with annual funds at its disposal to subsidize city centers of arts and drama, be set up struck us as a more practical and fitting tribute than lifeless and unconstructive war memorials and statues.

—SGT. JOE EGER of the U.S. Army Air Forces Band,
and EDWARD WM. ARIAN, Navy Instructor

When Ed and I were students prior to volunteering for war service, we had helped to create an organization of students and young people to promote the arts. We called it Youth Arts Forum (YAF) and talked about the need for national support for the arts. After the war, Salkind went to Juilliard to become a pianist, and he eventually became president of the San Francisco Conservatory of Music. Ed joined the Philadelphia Orchestra, became a distinguished Ph.D. professor, wrote books, and was chairman of the Pennsylvania State Council on the Arts. But in those days we were just GIs (government issue), a flexible term that could describe everything from military garb to a rigid, "just follow the rules" soldier.

Soon after the letter was published, Josh Billings, a prominent figure in the arts, invited me to speak in New York's Town Hall in January 1946, before a daunting group of Broadway and film actors, directors, dancers, visual artists, musicians, and others in the New York arts scene. Sponsored by the Independent Citizens Committee of the Arts, Sciences and Professions, ICCASP subsequently produced a bill "For the Establishment of a National Cultural Program," which was a forerunner to our present National Endowment for the Arts.

The idea of honoring victims of tragic events by initiating new cultural programs is still an issue today. A letter published in *The New York Times* suggested the best use of the World Trade Center site:

> Officials should consider creating an international cultural center for the study of tolerance, diversity and humanity . . . dedicated to the victims of terrorism and to the heroes. . . . Buildings would house libraries, research facilities and open studios for scholars and artists. Fellowships could be established to foster original work by an international community of scholars and experimenters who choose to pursue a new perspective or who want to preserve and learn from an ancient culture. Performance and exhibition spaces would attract visitors to a sanctuary dedicated to what humanity can be.
>
> —VALERIE HUNT STINGER, Palo Alto, CA

Eleanor Roosevelt was a good friend of the arts and of artists. When the great African-American contralto Marian Anderson was refused a stage in Washington, Mrs. Roosevelt came to the rescue in a landmark concert she hosted at the front of the Lincoln monument. Her support of minorities and the poor never flagged. Among her many friends was one who shared her social outlook, Albert Einstein.

53

Artistic Freedom

"I write [music] as a sow piddles."
—WOLFGANG AMADEUS MOZART

"Sensation," an exhibition of British contemporary art drawn from the collection of Charles Saatchi, the London advertising magnate, pervaded the atmosphere a few years ago. Mayor Rudolph W. Giuliani and Roman Catholic leaders zoned in on one artwork, which depicted the Virgin Mary decorated with elephant dung. Senators used this bit of cynical poor taste to criticize the entire Brooklyn Museum—then staging the exhibition—and threatened to cut off financial support. Right-wing foundations joined the snarling pack, and the very existence of the venerable museum was in jeopardy. The next step was to attack the NEA. Funding of all museums was threatened, and the incident provided a convenient excuse to cut the NEA's allocation and replace its leadership with conservative friends of prominent Republican senators. Intimidated, the American Association of Museums announced at the July 13, 2000, association board meeting that it adopted new ethical guidelines on how museums should oversee displays of art. The effect was to frighten contributors and expand an insidious climate of self-censorship in *all* the arts, including television and the news media, which, after September 11, fell into lockstep with the government's spoonfed news.

Apparently, the association, museum heads, and government officials realize the enormous power of the arts and are trying to curb arts freedom.

The Science Budget

As in the arts, the U.S. Congress cut the budget for basic science research. This has gone on every year since the 1970s. As a result, fewer and fewer Americans are entering the field of physics. Alan Chodos, senior research physicist at Yale University, wrote in *The New York Times:* "Any American undergraduate who considers physics as a career soon realizes that after finishing 5 or 6 years of tough graduate work he or she will be faced with a tight job market in which low salaries are the norm. . . . This year's eight theoretical physics graduate students include a Russian, a Georgian [from the former Soviet republic], two Indians, two Chinese and a Greek. And, oh yes, one American." (To make things even worse, the current fear tactics keep out foreign talent.)

Chodos concludes: "Congress seems inclined to address the problem with calls for heads to roll and security to be tightened. Some of these measures are not only unnecessary, but we also need to approximately double the number of Americans who are being trained at the graduate level in the basic sciences. After years of erosion and neglect, this can be done only by providing the resources to restore the image of the profession. One hopes that when the current frenzy of fear, loathing and political recrimination subsides, the serious work of rebuilding American science will begin. . . . it is not that we must guard against foreign nationals who are conspiring to infiltrate our national laboratories. Rather, it's that American scientific pre-eminence is at risk because there are so few good, young American physicists."

Science is increasingly focused on those subjects that have "practical" use for the pharmaceutical industry or for "defense." Politicians and some university heads want to make sure that science and the arts serve this particular purpose and no other. Seeing fat contracts, universities vie for the privilege of selling their technical expertise and theoretical knowledge to multinational firms. Thus does academia supply a safety facade and academic legitimacy for the huge companies who sell pharmaceuticals and agricultural products.

In education, our schools emphasize dates, standardized tests, and packaged interpretations of history, all of which leave little room for creative and critical thinking and play. Drastic cuts in Head Start and food programs leave thousands of children behind, separating the poor child from the rich. An empty belly doesn't digest knowledge. Television and other media add to the narcotizing of children and adults. One would have other expectations of scientists, but they too are locked into a system where fragmentation is the norm. Their brilliance and talent is spent in some tiny area of research or in creating ingenious television commercials. This is not to put down the value of specialists working to discover a small piece of the puzzle that when fit into other pieces may reveal a scientific breakthrough. But the majority are involved in research serving military or commercial interests. Consequently, they are dependent on government and corporations for their livelihoods. Those exploring areas of basic research are poorly funded, if not completely neglected.

And How About Music?

> "Style is [nothing more than] getting from one note to the next."
> —DIZZY GILLESPIE

Like science, music and its style systems are merely artificial constructs developed by musicians within a specific culture. Music's "basic research" is simply a series of variations based on "inventions" and previously developed "discoveries." The existence of many different musical style systems, both in different cultures and within a single culture, demonstrates that musicians construct styles in a particular time and place, under specific conditions. Styles and genres are not *completely* based on universal relationships inherent in the tonal material itself but are acquired through learning that comes within a specific cultural climate. In that sense alone, the experience of music is not a "universal language." The languages and dialects of music are many, and though they vary from culture to culture and from epoch to epoch, they are still dependent on—even in the departures from—the frequencies of the harmonic series. Many devotees of "serious" or pop music have difficulty enjoying jazz, and the same can be said about jazz aficionados. Recognizing the diversity of musical languages, we must also admit that these languages have important characteristics in common.

Certain musical relationships appear to be well nigh universal. In almost all cultures, for example, the octave, fifth, and fourth are treated as stable, focal tones toward which other notes tend to move.

For example, in the sixteen introductory measures of our old friend, Beethoven's Ninth Symphony, the harmonic structure provides the lattice upon which to hang the ensuing melodies. The triad is the operational norm. The open fifths in the introduction give a sense of incompleteness, seeming to presage some momentous event.

The sense of relentless power, which characterizes the main theme of the first movement, is a result not only of the elemental force of the theme itself and of the ambiguities and expectations excited by the introduction but also of the particular manner in which the introduction leads into the theme. This one does not conclude with a dominant preparation, a waiting period in which the listener is given an opportunity to orient himself to what has passed and prepare for what is to come. Instead the music moves on without pause, without pity, to its awesome declaration.

Beethoven's music, mirroring the essential shape and substance of human experience, from time to time contains sudden, shocking clashes with unpredictable chance. Lesser composers tend to eschew such harsh encounters with the unexpected, avoiding them by employing a single-minded sameness of musical materials or minimizing them by making a fetish of well-oiled, smooth transitions. But the greatest masters have faced fate boldly, and capricious clashes with chance are present in much of their finest music.

Both music and life are experienced as dynamic processes of growth and decay, activity and rest, tension and release. These processes are differentiated not only by the course and shape of the motions involved in them but also by the quality of the motion. A motion may be fast or slow, calm or violent, continuous or sporadic, precisely articulated or vague in outline.

Thus, particular musical devices, such as melodic figures, harmonic progressions, or rhythmic relationships, are formulas that indicate a culturally codified mood or sentiment.

In a series of children's concerts I gave at the New York Historical Museum, at Carnegie Hall, and in the schools, I would ask the children to imagine, as they listen to a pasage of music what the composer is trying to say, what the music reminds them of, or how it makes them feel. I would determine the length of the passage by the age of the children and their collective attention span.

Their responses reflected their family background, what they had read, what was happening in school or in the world. A boy said it made him want to dance. A little girl said one part made her feel sad. One boy thought it was about dragons. All the children in every concert seemed very happy to be invited to participate—as creative listener/participants. They were being acknowledged. Adults were respecting their views and feelings.

When the orchestra players demonstrated their instruments, the children described the evocations from the instruments. The bassoon sounded "funny" and the French horn "sad." The piccolo reminded one little listener of a July Fourth parade marching band. The tuba was a comical monster. Music became less abstract and had entered their lives.

54

Music: Agent of Social Change

In the seventies, I produced a series of three-day con-
cert/seminars titled "The Power of Music: Instrument
of Social Change," first at the Smithsonian Institution
and Woodrow Wilson International Center in Washington, D.C., then at the venerable Franklin Science Institute and the Museum of Philadelphia, at New York's
Columbia University, and two weeks for the Salzburg
Seminar in Austria. These concert/seminars combined
discussions with performances of music, dance, and
exhibitions of visual artworks. They attracted artists,
businessmen, musicians, doctors, and professionals.
Librarian of Congress James Billington, head of the
Smithsonian Institution, was our first host. His remarks
included his sense of the insufficient appreciation of the
power and precise role of music in the development
of modern human culture, and an exploration of the

creative dimensions of human imagination, including a discussion of "vertical montage," the juxtaposition of sounds to visual image in film (developed accidentally by Prokofiev when working with filmmaker Sergey Eisenstein), as distinguished from the mere succession of visual images. The remarks of this distinguished cultural historian are worth repeating:

> As words become more and more a subject of division, confusion and profusion, the unifying forms of music are of increasing importance. During . . . the film *Alexander Nevsky,* one becomes deeply emotional during the famous charge scene merely on the strength of the musical score. The whole effect is far more profound because it's been musically motivated than it would have been if it had been simply pictorial.
>
> At the time of their death both Prokofiev and Eisenstein were actually on the verge of a breakthrough in the use of music rather than the visual image, which is relatively limited in its potential and relatively defined by its inflexible interpretational quality.
>
> Similarly, the people who developed the study of language in the early 20th century turned to music. One of the great stimuli in Russia was the discovery of a kind of Rosetta stone for reading pre-Byzantine Russian native music. They discovered a set of equivalences showing that there was a multiplicity of musical languages yet to be discovered by the archeology of the future.
>
> Two of the most important revolutions were started by musical performances. You would wonder whether it's the impact of "La Marseillaise" or of certain operas (now mostly witnessed by people who can afford to pay high prices). How did these songs and operas actually drive people to revolution in a way in which people kept expecting Woodstock to do and never did?
>
> Whether it is "Ein Feste Burg" on the Reformation or the "Marseillaise" on the French Revolution or, for that matter, the musical score of *The Triumph of Will,* music is a powerful, unrecognized force in the whole dynamics of human communication and motivation.

This is a plea for a study of music and appreciation of music in history and general cultural studies. We haven't been willing to recognize the extent to which the transverbal motivators have really driven people and have determined not just what people are going to do in a physiological sense but the inner definition of peoples' feelings, commitment and beliefs.

The secrets of music and its evocative powers, whether in the hands of a demon or in the hands of a redeemer, are going to play a role, as they have so often done in milestone points of the historical past.

First Wave of the Revolution: Rock Music

In the sixties, the exponentially greater audiences that developed from exposure to records, radio, and television were beginning a transformation from listener to performer. Young people wanted a piece of the action. As great heroes and leaders fell by the wayside from assassins' bullets, corruption, fraud, and a widespread betrayal of the public trust by entire government administrations, a whole new younger generation shifted its trust to the old verities: individual conscience and the arts, and especially to *their* music—rock. The antecedents were the blues, African-American rhythms, Broadway, and jazz, now aided and abetted by the new electronics. The young of all races were searching for their own voice; music was it—at once a bloodless revolution and a dream of a better world, as in the Lennon song, "Imagine."

At one time, the dream of a socialist or communist utopia promised the path to an idealized human existence. Then the reality became misery and strife. The experiment almost died with the dissolution of the Soviet Union. Dreams, however, are not that easy to totally extinguish. The embers began to glow in new forms and mechanisms for information transfer, enabling discussion and creative activity outside the conventional media. Technologies facilitated change.

The Internet began to provide an organizing tool for huge numbers of people to communicate instantly with one another. A call to action gets quick results, as witness the quickly organized antiwar demonstrations and Howard Dean's presidential campaign, built largely on the Internet. A new sensibility has grown up among millions. The mood of the formerly passive listener changed as there began an incipient demand for active participation in the governing of their lives. For the young, music was still in the vanguard. The instrument playing this revolutionary music was the computer.

A New Wave

The implications have not yet clearly emerged, but the general trend is unmistakable. There is no blueprint or tightly woven score, but the notes multiply: ethnic associations, women's liberation movements in unlikely countries, consumer and tenant groups, et cetera. A new energy continues to seethe beneath the audiences of people who turn out for concerts, exhibitions, festivals, and those who access the World Wide Web.

Though string theory points to a musical harmony as the basis for all phenomena in the universe, and physics has given us insights and processes that opened the door to the greatest advances in human history, a tragic vacuum is yet to be filled—the scientific and humane approach to human society.

Music and physics point toward the recognition that we are all in this together despite the formidable forces mobilized to separate us. A few giant corporations own or control education, communication, the food we eat, the air we breathe, and the water we drink. They own radio, television, the press, the armaments industry, and the government. They plunge us into bloody wars for profit and power. What force could turn around this powerful engine steamrolling downhill, gaining momentum by the second? Is "the power of the people" just an old-fashioned cliché?

Music and musicians have constituencies far greater and more passionate than any president or head of state. Recordings of dissident musicians have sold as never before. Books on physics and antiwar and satiric diatribes against lies and corruption gain the bestseller lists. Scientists, if they get their act together as they did in creating the atom bomb during World War II, could control all of technology. The stakes are higher today—the survival of life itself. President Eisenhower's warning about the dangers of the "military/industrial complex" was never truer, and the awareness of these dangers was never greater. If we can unite electricity and magnetism, the quark and the big bang, we should be able to unite peoples. Musicians and scientists can make a signal contribution toward educating and uniting the population. When facts and truth are smudged beyond all recognition, education is one of the main vehicles to carry us toward Beethoven's "brotherhood" of all humankind. He would have no trouble adopting that cliché, "the power of the people." Nor would Jesus. People are becoming disenchanted with government but are skeptical of all politicians and leaders. Can physics or music, as defined in these pages, help light the path to a saner and more compassionate world, saving the world from its deepening crisis?

55

The Crisis Deepens

Today's crisis in physics and in music is a reflection of the many other crises affecting the world today. This crisis goes beyond the contradictions between a Newtonian worldview and the new physics. Nor can responsibility be laid at the doorstep of quantum mechanics for some of the lunatic and perilous schemes, theories, and practices hiding behind the shield of the new physics. Religious fundamentalists, big business, and politicians, especially of the neoconservative variety, have been quick to appropriate quantum mechanics and a perversion of the new music to sell their fundamentalist religion, anti-Darwin ideologies, and biological nightmares.

Television, drugs, guns, violence, and an erosion of values—"family" (whatever that means) or otherwise—add to the mix. All these are only symptoms of a deeper malaise. They are to be deplored, transmogrified, and excised. But like cancer, they will quickly metastasize unless they are reached to their foundations, to the cause or causes, in order to find the elusive cure. An in-depth study of the elements that contributed to the crisis is in order. Republicans, Democrats, corporate CEOs, chemical in-dustries, and greed are no doubt the prime villains. We can even add Osama and Saddam, but we must still dig deeper. Could it be capitalism itself, the attendant consumerism, and all values subservient to the drive for ever-greater profits?

It is a maxim that at every stage of the observation of nature, latent contradictions come to light. The contradictions between the Galilean laws of motion and Kepler's laws of planetary motion led to the New-tonian equation of mass to inertia and the formation of the universal law of gravity. Newtonian science gave us many conveniences (washing ma-chines and vacuum cleaners) and questionable benefits (SUVs, snowmo-biles, and water scooters). Newtonian science lifted our physical and material world to a higher level, at least for many. But the social/spiritual world of human relationships has suffered by an across-the-board New-tonian approach of separation and division.

Our faith in technology is unwavering; we are buried in gadgets, toys, and computer games. Our government allocates astronomic sums, with very few dissenting votes in Congress, to ever more sophisticated weapons that can do ever more damage per square mile, and to weapons, contracted to the overblown "defense" contractors, that become almost immediately useless in the face of changed tactics in warfare waged by individuals instead of tanks and planes. Huge subsidies go to rich, cor-porate "farmers" to grow more and more manufactured, chemical, and possibly toxic food. We build better prisons and stock them with racial minorities who might vote against or rise up against the privileged few. The United States has the largest prison system in the world, incarcerat-ing more than 2 million more people than all those in the rest of the world combined. One of every 143 U.S. residents is in federal, state, or local custody—half of them for nonviolent crimes, often petty larceny, that have mandatory minimum sentences as much as thirty years to life. When finally released, they can't vote, can't get jobs, have no skills, have no place to live, can't get driver's licenses or student loans. Many end up back in prison.

Warehouses and markets are glutted with "stuff," which diminishing numbers of buyers are able to afford. Inventories are either over- or understocked. A billionaire elite looks with fear at the great middle class, who have been manipulated at elections and now worry about unemployment and poverty. Some of this great American middle is awakening to the fact that they sold their birthright. They helped build a political hierarchy, which, on taking office, further enriched the already corporate rich and expanded an already bloated military machine to protect this elite from everyone else in the world. Children are turned into educated automatons from controlled curricula and consumers of commercial product, if not sent to war. Schools vie for the alms given by Coke and Pepsi, for the right to sell their addictive sugared water to the children, already overweight from junk food. A blanket of infotainment in television and the print media dumbs down the public. The arts in education are cut or choked to death by lack of funds.

It would be a great mistake, however, to lay all the ills at technology's door.

Technology in Music

Technology, in the right hands, in a rational, scientific societal system, could free the peoples of the world from want, repression, and misery. To start with, let's look at where musicians and physicists fit into that dire-cum-rosy picture.

Steve Reich, whose compositions have achieved more public performances and acclaim than most contemporary composers, has just written an opera with his video-artist wife, Beryl Korot. *Three Tales* expresses his fears about the perils and misuses of modern technology. This is not his first musical/theatrical composition that focuses on an important subject. He also composed a video-opera meditation on the Holocaust, *The Cave,* in which Palestinian Muslims, Israeli Jews, and

Americans discuss the significance of Abraham, with live musicians playing "speech melody." Reich is one of several leading modern composers with a social conscience, as is explicitly evidenced in his book *Writings on Music.* Philip Glass and John Adams have been satisfied to be more implicit in their messages, as in Glass's *Gandhi* and Adams's *Nixon in China.* The Rolling Stones, for their part, are collaborating with the National Resources Defense Council to raise awareness of and fight against global warming. And Bruce Springsteen and other famous rock performers toured the nation in 2004 in an effort to unseat George W. Bush.

Rock, pop, and folk composers have long ridden Paul Revere's horse. Another rebel to win the hearts of massive audiences from youngsters to aging boomers is twenty-two-year-old Conor Oberst who hisses his angry lyrics: "They say they don't know when, but a day is gonna come / when there won't be a moon and there won't be a sun. It will just go black. . . . It's hard to ignore all the news reports. / They say we must defend ourselves. / Fight on foreign soil / Against the infidels / With the oil wells. / God save gas prices." He says, "Political songs [are] the only thing worth singing about. It's difficult to think about anything besides the war." Oberst asks himself, "Am I pure?" His young fans understand and respond to this self-probing integrity.

In the early forties when I was studying at the Curtis Institute of Music, a constant visitor to the one-room pad that I shared with three other students was an accomplished composer, Earl Robinson, well loved in the politically progressive, folkish world of the time, but hardly mainstream. He had composed *The Lonesome Train,* a funeral cantata after the death of Abraham Lincoln, and *The House I Live In,* a Walt Whitmanesque paean to *What Is America to Me.* Some generations later this song was picked up by the Republican Party for their theme song at their convention, and was sung frequently by Frank Sinatra! Earl was highly amused by this mainstream recognition, although it arrived late in his life.

Ronald Reagan, in his 1984 reelection campaign, appealing to the patriotic sensibilities of young people, mentioned the song "Born in the U.S.A." by Bruce Springsteen. The problem with his reference was that the song is hardly one of jingo patriotism: "Sent me off to a foreign land /

to go and kill a yellow man. / I was born in the U.S.A." Similarly, George W. Bush asked Tom Petty for the use of his song "Won't Back Down." The epitome of counterculture, Petty was outraged at the request. It was doubtful that either Reagan or Bush had ever heard these songs; in any case, the message was certainly lost on them.

Musicians, artists, and yes, scientists, have an estimable history of fighting for ideas, for truth and principle, sometimes at risk of their lives and livelihoods. Scientists are making their voices heard in full-page ads about global warming and genetically altered food.

A Quantum Cure?

The postmodernist alternative to our modernized, mechanistic society posits an ahistorical, irrational, and chance universe in which anything can and does happen. It assumes that God's job is to roll the dice. The quantum "revolution" gave birth to the most cockamamy theories, couched in highly philosophical and obscure lingo. Even some well-known scientists bought into this chimera. Papers were read in academic circles, pundits received lucrative grants from right-wing foundations and won huge television audiences on all the major networks, and erstwhile criminals such as Oliver North were given their own television shows. Small cable stations were squeezed out as even National Public Radio joined the omnivorous pack, hoping to save itself after being attacked as "too liberal." Good music stations were either cut out or forced to join the Top 10 and pop stations. Fewer and fewer music classes in the schools remained alive.

These diabolical events did not spare scientists and musicians. Some professors and entertainers played the shill, earning their economic place in the sun while many of their colleagues, with notable exceptions, were afraid to talk up and lose their temporary sinecures. As my mother might have said, "Constructionists, deconstructionists, be *gesund*," which, roughly translated, means "the main thing you should be healthy." Not a bad take on the stratospheric debate among academic intellectuals. Some computer scientists came down to earth by creating a high-tech hand-held device that gives concertgoers a blow-by-blow "explanation" of the music as they listen/watch/learn/live in virtual simultaneity.

The Backlash

But all is not lost. Students, minorities, artists, entertainers, scientists, and businessmen (some who want the business from Cuba!) have joined a backlash. A new awareness grows out of personal interests and experiences. Even conservatives have become aware that their Constitutional and electoral rights are at risk, as are their children's futures. Political compromises, obtained at the point of a gun, can return to haunt the compromiser as well as the gunslinger.

It may be useful to remind us at this juncture of the definition of the word "crisis." We know what we mean when we say a business crisis or a political crisis. In literature, theater, and music, a crisis is the high point in a story or composition, at which hostile elements are opposed to each other. In medicine, it is the stage in a disease when a decisive turning point occurs, leading either to recovery or death. We certainly seem to be hurtling toward the latter occurrence at rocket speed.

Where Did It Start?

Students of history speculate that in ancient times, prior to civilization, human beings lived in harmony with "nature" and with one another. To put it in holographic language, the whole lived in each part and each part was integral to the whole. Four or five thousand years ago, things started to change, but not for the better. People began to compete with and become separated from one another and from their environment. Nature was something to be conquered and exploited, as is explicitly commanded in the Old Testament. Wars, crime, slavery, selfishness, and injustice crept onstage. The Greeks applied a "civilized" facade, and things continued downhill with only a few temporary impediments, such as the eighteenth-century revolutions in France and the U.S., the Enlightenment, and the twentieth-century social improvements, such as the U.N., human rights, and Social Security. Now even these bones are being subverted.

The germs of deterioration are growing to epidemic proportions and infecting the entire world. This crisis is of such magnitude that it can be solved on a stage no smaller than the world itself. No one discipline or government, even the imperial one, can solve it. The international World Health Organization, despite its earnest efforts and praiseworthy accomplishments, is unable to heal sickness on any large scale, nor can Alan Greenspan, head of the Federal Reserve Board, solve the U.S.'s or the world's economic crises.

As the downhill pace quickens, the legal pirates realize that their time may be short, so they gobble up all they can before being exposed. Their clever public relations gimmicks become more obvious daily, for the simple reason that they contradict peoples' actual experiences. The terrorism of poverty, hunger, and disease, all unnecessary in a rich, technologically advanced world, nourish hate and resentment worldwide. Unfortunately, some of this hate becomes misdirected as artificial divisions cause strife between groups instead of uniting in defense against global corporatism. "Divide and conquer" has always been the watchword of power elites.

The crisis can be treated successfully only at its causal base. All else will be cosmetic. It is a crisis of society as a whole.

56

Our Post-Human Future

Arundhati Roy was asked by her foreign friends why she doesn't leave New Delhi. Doesn't she think the threat of nuclear war is real? "It is," she answered, "but where shall we go? If I go away and everything and everyone—every friend, every tree, every home, every dog, squirrel, and bird that I have known and loved—is incinerated, how shall I live on? Who shall I love, and who will love me back?"

She and her friends decided to stay. "We huddle together. We realize how much we love each other, and we think what a shame it would be to die now. The macabre has become normal. While we wait for rain, for football, for justice, the old generals and the eager boy-anchors on TV talk of first-strike and second-strike capabilities as though they are discussing a family board game. My friends and I discuss *Prophecy,* the documentary about the bombing of Hiroshima and Nagasaki . . . the dead bodies choking the river, the living stripped of skin and hair. . . . We remember especially the man who just melted in the steps of a building, we imagine ourselves like that. As stains on staircases. That's what nuclear bombs do, whether they are used or not, they violate everything that is humane. They alter the meaning of life itself."

Physicists in the smallest nations understand the principles of constructing a nuclear bomb. The only thing preventing them from doing so is the cost of production. A megaton bomb can produce the explosive force of a million tons of TNT. The original fission bombs dropped on Japan have one-fiftieth the explosive force of a megaton bomb. One such bomb would destroy the entire population of New York City and environs. The delayed and long-term effects are impossible to estimate. The Chernobyl accident still causes children to contract leukemia, cancers, and genetic defects in the Ukraine and in other countries, depending on the wind direction. The U.S. is radiating whole areas of our country with the toxic garbage from bombs that have a half-life of tens of thousands of years. If, as we are told, they are being safely buried, why then do governors (mainly Republican) fight against having them buried in their home states!

Leon R. Kass, appointed by President Bush to chair the Council on Bioethics, was charged with recommending federal policy on cloning and stem-cell research. Kass has both an M.D. from the University of Chicago and a doctorate in biochemistry from Harvard. He was a founder of the Hastings Center, the first American bioethics think tank. His concern over the out-of-control technology arises from the very nature of what it means to be human. In his *Life, Liberty and the Defense of Dignity: The Challenge for Bioethics,* he uses technological and bioethical concerns through the lenses of the "morality . . . love and creativity . . . that characterize our species." Kass worries that once scientists have begun to produce cloned embryos, it would be virtually impossible to control what was done with them. In the current political/social context wherein anything goes when it comes to turning a dollar, Kass has a point.

All technology can be thought of as biotechnology, since technology boils down to machines interacting with humans. According to Francis Fukuyama, in his *Our Posthuman Future: Consequences of the Biotechnology Revolution,* "Technology is designed to overcome the frailties and limitations of human beings in a state of nature—to make us faster, stronger, longer-lived, smarter, and happier." But he asks: "Are our lives really better for the existence of the automobile, television, and nuclear power? These questions are ethical and political, philosophical and spiritual as well as medical. There are doubts about whether the human soul thrives best in the technological world we have created for ourselves." Who wouldn't ruefully concur when Fukuyama says: "I am continually struck by how much time I have to spend fixing the machines that supposedly improve my life."

The Luddites in early-nineteenth-century England wanted to destroy the machinery that threatened to take away their jobs. We can't return to pretechnological times, but perhaps technology could serve humanity once capitalist exploiters are made to release their iron grip on these tools and the tools are wrested from the hands of those who use them only to gain profits and maintain control over the people who made them.

Today we face a new phase in the powers of human technology as we discover and implement an alternative to evolution itself—direct intervention in the genetic process. We create animal clones and argue about extending that to humans, even discussing what "desirable" characteristics we should implant. Long past the previous generation's fascination with artificial intelligence, we talk instead of neural implants to enhance human abilities. We can ask: Abilities for what? To make more gadgets? More money? To vote "correctly"? Who is to decide which abilities are favorable and which are not?

We are flouting the natural order and nature's timing. It took thousands, indeed millions, of years to accomplish evolutionary changes. The result of compressing these changes into a few years or months is frighteningly evident. Of global weather changes, warming is only one. At this writing (summer 2003), the temperature in New York is 105 degrees Fahrenheit, and much of the country is suffering under similar heat. Meanwhile, despite conclusive evidence from almost all of the scientific world that global warming is happening, the Bush administration pooh-poohs the idea that emissions from chemical and industrial sources that his administration has deregulated is causing anything abnormal. The unknown consequences of chemicals in air, in water, and of genetically altered foods are unimaginably scary!

A toxic environment can corrupt and sicken babies, those divine creatures who, not born evil, become so from poor nutrition and polluted water. These pernicious influences invade the minds of children in the schools, where overloaded teachers are restricted to teaching by rote from prepackaged government and officially altered textbooks. And the separation of church and business from the classroom has gone the way of the dodo.

57

Science and Religion

The American president also brought to the forefront the debate between science and religion. For a time in the past century, it seemed that science was winning. But when George W. Bush and his conservative team came to power, it brought the debate "in your face," front and center. The debate entered every corner of American life, including letters to the editor and op-ed perorations, and in churches, synagogues, and institutes. The Florida Key West Literary Seminar of top scientists and science writers, intended to celebrate the triumph of science, found themselves mainly discussing religion. The consensus was that theories such as the big bang, and evolution have displaced humans from the center of the cosmos and made God less necessary and less likely. Nobel laureate Steven Weinberg stated flatly, "Religion is largely wrong." Some panelists insisted that the yearning for meaning, expressed commonly through religion, is an irresistible part of what makes us human. "We are the products of millions of years of evolution," said Lynn Margulis, a geneticist and author. "But we believe in lies and stories, and that's what makes us unique. For writers, the task is to bring all of our feelings, including religious feeling, to inhabit our thoughts of science. I think there is room in our view of science for religious feeling." Another Nobel physicist, Murray Gell-Mann, saw no reason to attach the word "religion" to any of the good things in science or in life, such as love, compassion, or other attributes traditionally thought of as the province of religion. "Why let religious people have the good things?" he asked.

Weinberg responded: "Perhaps the word 'spiritual' would be better than 'religious,'" though he noted that concepts of spirituality are tainted by superstition. He also noted that science cannot inspire the artistic imagination in the way that religion produced some of the great music, sculpture, and painting, from Bach, Dante, and others commissioned by the Church. Weinberg added that the secular imagination is superior to both science and religion, and added, "I do believe in tolerance. I should be nice to religious people, and they certainly should be nice to me."

An interesting side note has been the attempt by astrologers to heal the schism between science and religion. To be sure, there has been a fraudulent aspect to astrology; nonetheless, it attempts to see the world as a living realm, to find some deeper purpose in existence itself. Admittedly, the new-age astrologers distort some of the findings of quantum mechanics, but let us also admit the distortions by some scientists. There is an awesome world out there with mysteries far beyond our current knowledge. Surely if our relatively small neighbor the moon can move the oceans back and forth, how much influence on us might the sun and the stars have? Astrological predictions are not far off from some of the "far out" theories authored by scientists.

The Gods Must Be Crazy

In the film of this name, a small airplane flies over a remote area in the Kalahari Desert where no plane had ever flown. The pilot takes a sip of Coca-Cola and throws the empty bottle out of his plane. A man on the ground who has been watching this loud bird in the sky sees the Coke bottle dropping from the heavens and approaches the strange object. Curious, he picks it up and tries to blow through it. Managing to get a sound, he takes the bottle back to his tribe, where he describes the strange incident.

The tribesmen conclude that the gods must be crazy. Gradually, however, they begin to think that this bottle (now a holy object) might be some kind of emissary from heaven, and start worshipping it.

Today's schoolchildren seem to worship the same Coke. We too worship the technology that brought us the internal combustion engine, air-conditioning, the electric light, and the computer chip.

Sure, we all thank the gods for the electric light, but I'm not so sure that the devil wasn't whispering sweet nothings into the ears of the gods when they gave us DDT, Agent Orange, nuclear bombs, chemically resistant mosquitoes, and my computer! Some might want to throw in a few other mixed blessings to send back to the gods—plastic, supersonic airplanes, nuclear waste, and SUVs. Our wondrous technology can be employed either to destroy or to bring undreamed of benefits.

Not long ago, science and religion barely tolerated each other. But this has been changing, especially after the theories of quantum mechanics entered the scene. Religious leaders, realizing the hold science has on the public, have made accommodations, turning science inside out to support their belief systems. After all, they argue, who created the big bang if not God? Fundamentalists who insist on the verbatim interpretation of the Bible (Adam and Eve, the apple, etc.) have gained enthusiastic audiences and enormous subsidies from foundations and government via faith (Christian) based funding to schools, prisons, and television. Somehow, some scientists even see confirmation in religious dogma of their views about quantum mechanics, just as religion enlists science to give it academic credentials. Quantum mechanics is a boon for those whose interests lie in making the link between religion and science.

To be sure, who among us has not been awed by the complexity and beauty of a flower, a mountain, the ocean, a newborn life? These and other wonders of nature have never ceased to astound us. While we unleash the power of the atom and open doors to the wonderful and terrifying forces in nature, we continue to ponder the meaning of life: Why are we here? Are we born inherently and inescapably evil, doomed to carry the baggage of original sin forever?

How important is man? Religion puts man on top of the heap. Scientists see him as insignificant in the cosmic perspective. The story of creation is just a story to most scientists, though it is fact to fundamentalist religionists and a tiny few religious scientists. The word "scientific" usually connotes that which is measurable and provable. Mysticism or religion demands no such proof. Faith is enough. Einstein took the position that pure scientific research was based on "objective, causal reality"; in other words, the scientist observed cause and effect, testing his observations separate from himself. On this, Neils Bohr and other proponents of quantum theory questioned Einstein. Though Bohr, on the whole, agreed with Einstein, he argued that experiments that led to quantum theory changed the outlook of physics and the very laws of nature, which are ultimately only probabilistic.

Quantum mechanics led to these questions: Is there any reality in what we perceive? Do we really affect the so-called reality that we see by simply looking at it? We have learned that quantum theories describe a reality that is dependent on the observer. Are we then playing God by creating our own reality? One can understand the confusion that quantum research led to in the philosophical, religious, and scientific worlds. A Boston Red Sox fan doesn't watch their games because, he says, his presence puts a jinx on them! After all, quantum theory has correctly predicted atomic spectra, superconductivity, lasers, antimatter, and other phenomena.

The phenomenon of photons "communicating" with one another could be interpreted to mean that communication travels faster than the speed of light, an impossibility according to relativity theories, which have *also* been proven over and over again. If the photons are sufficiently far apart, then the distance could not be traveled, even at the speed of light! Thus the implications led to predictions that have further obscured the difference between science and mysticism.

Quantum mechanics has led to fantastic science fiction and has been seriously argued among scientists and in books such as *The Tao of Physics* by Fritjhof Capra, *The Dancing Wu Li Masters* by Gary Zukav, *The Matter Myth* by Paul Davies and John Gribbin, and *The Scientific Search for the Soul* by Frances Crick.

58

God Gets into the Ring
with Causality

I f one reverts to the etymology of "religion" *ligare,* which means to bind together or connect to some superhuman power, and *religio,* which is reverence for these powers, then history could parallel the development of religion. Connections are what religion is about, connection to the All (holiness or the whole), also translated as God or Nature.

The mainly positive effect of religion and music on and for each other has been clearly established for centuries—but less positively for science. In Newton's time, the bourgeoisie advanced against feudalism by backing a Protestant God against a Catholic one. Toward the latter half of the eighteenth century, religion allied itself with the landed aristocracy and finance capital. Revolutionaries such as Voltaire and Lemaître stripped matter of as much of God as possible and became atheists. According to Kant and Hume, the mind of the individual replaces God so that determinism is drawn out of experience.

Determinism (which says that all life is predetermined, even one's smallest day-to-day decisions) and causality (a cycle of cause and effect) have seesawed back and forth. Philosophers and scientists succeeded only in continuing the confusion and uncertainties. Fundamental religionists, however, are sure they are right, others wrong. Jewish, Christian, and Islamic fundamentalists, unfazed by nuanced interpretations, know with absolute certainty what is God's Word. Over the centuries they've drawn tons of blood to bolster their positions.

Poor God must be very busy keeping up with all the contrasting interpretations of His words and laws. If God hasn't preplanned every single detail in advance to the forever (strict determinism), He then has to decide on each stock exchange transaction, each baseball game, each religion's interpretation of His manner of praying, every child's little white lies, and the infinite instances of other ungodly or godlike actions. God should go on strike for overtime pay. I'd like to suggest a modest proposal: that His vision of determinism versus causality might end up synthesizing the two into a kind of dialectical relationship, a marriage between free will and determinism, spirit and substance, objectivity and subjectivity, theory and practice, all in a state of permanent movement.

Causality

Toward the end of the nineteenth century, Newtonian mechanics became the model for all other sciences. Lord Kelvin argued that physics, in terms of a handful of physical laws, was nearing its end. According to Kelvin, the universe had been transformed from a living organism into something that is close to a machine, and its behavior could be reduced to an operation of its parts in motion. Even human nature could be a result of the flows of energy arising from electrochemical reactions of the nervous system.

Then came relativity and quantum theory, turning Newton's (or Kelvin's!) legacy on its head. Relativity began a process, climaxed by quantum theory, that transformed physics and challenged the Newtonian world and all associated with it. However, Einstein never lost sight of Newton's enormous contributions to science, giving this great scientist his due, for Newton's theories were and are still serviceable in our everyday macroworld.

Based largely on hunches and the notion of quanta put forth by Einstein and Max Planck, Niels Bohr claimed that quantum theory revealed the indivisibility of nature. Among Bohr's signal contributions to subatomic theory were relating the number and momentum of electrons in orbit around the nucleus, figuring out how the nucleus kept the electrons from flying out into space, and discovering the principle on how the electron radiated light—theorizing that the electron jumped from one place to another in orbit around the nucleus and that the energy released in this jump was emitted in the form of light. Bohr went on to become a mentor and adviser to many of the greatest scientists of the century involved in the creation of nuclear energy and the bomb. In 1945, secretive Winston Churchill persuaded Roosevelt to turn down Bohr's urgent suggestion to share the bomb's technology with the Soviets and other countries. This triggered the Cold War and a world threatened by countries small and large by proliferating nuclear bomb availability. Bohr believed that the world could take a lesson from physicists, who routinely share their discoveries to the benefit of everybody.

Separating effect from cause leads to mistakes in science and in human behavior. Let us take a homely example, the sound from a violin. The sound is traced by physics to a chain of cause and effect, which links the excitations of the listener's auditory nerve with compression waves in the air, these in turn with percussion waves in the wood "box," and this in turn to the relative motion of bow and string, and then to ingredients and the making of the fiddle and lastly to the player. Hence, in practice, causality is seen to be a very different conception from that of determinism. However, if we adopt the purely deterministic approach, the causal chain is incomplete and misleading. The violin and ear, being where they are, also depend on the location and movement of the earth. According to the principle of relativity, any change in one point in time would mean nothing else could remain precisely where it is. Moreover, the ear, air, and violin are all the product of a long process—going back to the creation of the universe—in which each stage has determined the following stage. Hence, everything in the universe is a "cause," a determining factor, not merely now but in the entire past. What the violin (or player) has been determines what it is now. One could logically say that all that is not ear-air-violin determines what *is* ear-air-violin. At first sight there seems to be a certain richness in this logical determinism compared to the mathematical bareness of scientific causality. But the ear-air-violin example is impregnated with activity. It stands out as an organized whole from the background of the universe. Something is "happening" against a background of "not happening." Wow! Are you as dizzy as I am?

Static Science and Music

Of course, there are happenings in the background, but they are dismissed because they do not participate in the same sphere of happening. So the scientist or the too strict follower of causality gets the colorless universe of s (sound) and not-s. As soon as the strict follower of determinism leaves the field of logic (equations, etc.) and enters the field of practice, he or she finds that determinism, as a strict definition, is self-contradictory. For example, if A B C D is generated according to the categories of determinism, there is a necessary connection between them. Given A, then B, C, and D must necessarily arise. The qualities of B, C, and D must already exist in A, and given B, A must necessarily have been its cause. And given C and B, A must necessarily be contained in it.

So much for strict, isolated causality or strict, isolated determinism. What we now have is a unity of opposites, a dialectic (thesis, antithesis, synthesis). Nothing can exist for itself. Time and space do not exist absolutely. They can never be separated. Nor can we (thank God).

Naive realism, or materialism, cannot be justified by theoretical arguments, only by practice. Life itself, in all its conditions, determines consciousness and beliefs.

Entropy

Eddington visualized the process of the universe as the gradual running down of a once fully wound universe. The world starts to die directly as it is born. But this precludes the principle of detailed balancing, that is, to every process there is a converse process, occurring with equal frequency. A man is born, a man dies. In using crowds of people, there are new births to counteract that individual entropy. Organizations of ants, people, or atoms are reduced to crowds of units. Statistics are all examples of particular events, and are ignorant of the processional relationships of other events that may soon contradict those statistics. The universe does not consist of separate particles in themselves but of particles *in relation* and in ever dynamic motion, intrinsically and extrinsically dependent on those processional relationships and events. So much for polls, political and otherwise. Thus the particle-in-itself involves uncertainty. The use of larger and larger numbers of particles involves greater certainty of probabilities. The only absolute certainty is whole universe (assuming there is only one!). The universe running down is also the universe evolving into new domains (order vs. new complexities). Entropy involves increasing disorder only.

This conundrum has spawned all kinds of speculations, such as many universes constantly being born and dying, and the reversal of time itself. I prefer to think of constancy of change, of continual process. Perhaps our sense of scale, both time and space, are still Newtonian. We place limits within our ability to measure. The context in which we view the universe is contained rather than free. We are the ants who can't imagine the immensity of the entire earth. Since the universe is made up of particles and domains, the quanta and the universe are both simultaneously certain and uncertain. What a riddle!

Freedom

This brings us back to causality and free will. Mechanism led to predeterminism, the idea that all history unrolls itself with iron precision. God controls everything from the very beginning to now and all time. In Newtonian, mechanistic science, the initial case had it that a giant calculator could predict the whole of the future with precision.

The uncertainty principle in physics, though interpreted otherwise by theocrats, gave scope for free will, freeing man from the nightmare of determinism. Man is now free to dominate and command nature to do what man wills. Man can now suck the earth of its life's blood to create commodities to sell in the "free" market. But the price is enormous. He has lost control of his social relationships.

When talking about the philosophical and scientific arguments over free will versus determinism, it might be useful to look into that much-vaunted "freedom" in the U.S. and the "free world." Young men are asked to give their lives in the cause of "freedom." We may be free to own a Jaguar and smoke expensive cigars, never mind the reality that most of the world's populace can't even count on the next meal. The son of a wealthy oil magnate has that freedom bestowed on him, earned or not, usually by consciously or unwittingly exploiting the labor of others. He takes the presidency of a great country as his birthright while attacking other countries for their failure to have "free" elections. The "free elections" gimmick, as just one more expensive commodity, afforded only by the very rich and big business, is a very handy tool to cloud over the subjugation of the world's people.

Until September 11, we took free speech, human rights, and freedom of religion for granted, at least in the abstract. But even before that infamous date, one had to be judicious about what one said in order to "get ahead." To adequately define the word "freedom," we must be conscious of causes and circumstances that unconsciously affect man's will. We all think we know what these words mean, but our error is rooted in our misunderstanding of history and the social genesis of freedom. We now relinquish our freedoms in the cause of war, of "defending our freedoms."

59

A Free Violinist and

a Free Atom

*"God tells me how the music should sound, but you
stand in the way."*
—conductor ARTURO TOSCANINI to a trumpet player

Unlike an atom, which is "free" to choose state A or state B, a truly free man can choose from a vast number of states. A violinist can "choose" between limitless positions on a string to achieve one set of string vibrations that sound in a desired pitch. Yet even this action has prior cause; his studies, his ability to pinpoint a desired pitch, and the "muscle memory" that enables his finger to press at precisely the right place on the string at the right time. Meanwhile, he may choose how much vibrato to use, depending on the desired musical result and the tradition of a particular musical passage. He may be constricted by the conductor's musical desires or by the composer's intentions. Yet his musical "freedom" is unquestioned.

The contemporary world, including scientists who should know better, tend to think of our minds and physical bodies as apart from nature; we "control" and manipulate nature. But nature fights back against this arrogant species with earthquakes, floods, hurricanes (four of which I recently experienced) fires, storms, sea changes, and more. We also have become apart from each other with the result that we are apart from ourselves, viewing our minds as separate from our bodies and the various parts of our bodies as separate from other parts. Our thinking and feelings are equally fragmented.

So what are the causes of terrorist actions, either by an Osama bin Laden or our own domestic terrorists like the Aryan Nations, Christian Identity, Posse Comitatus, and the like; or by the far greater terrorism, of states practicing wholesale, "collateral" killing of thousands of innocent people by remote control in sophisticated, often robotic planes? What do you call policies that subject millions of children to hunger, disease, poverty, and early death? Are these not also terrorism?

Is each action preordained, as some strict determinists say, as the prevalent attitude says: "It was always thus, always will be, and there's nothing we can do about it"? Or must we look back on the historical causes leading up to these actions? A causal approach might give us a more rational society, which in the long run could benefit everybody, including those responsible for egregious policies and dangerous programs.

Even scientists who have made incredible discoveries, such as the laws of thermodynamics, entropy, relativity, and quantum mechanics, have been bewildered by the contradictions inherent in the two approaches—determinism as against choice. Sir James Jeans in *The Mysterious Universe* has one answer: "The final truth about a phenomenon resides in the mathematical description of it; so long as there is no imperfection in this, our knowledge of the phenomenon is complete." Sounds like math is the physicists' God.

Einstein found and lived a solution to the enigma by keeping an open scientific mind while at the same time adopting an activist life of choice, choosing to make it a better world right here and now. Theory without practice is pure scholasticism.

Musical mathematics has attracted academically inclined composers, and some fascinating musical constructs have resulted, but for the majority of music lovers (including this professional musician), we want more human, more "user friendly" music.

60

Theory and Reality:
Does the Music Work?

What arrogance we humans exhibit! If reality is pure mind, then whose mind, or what combination of minds? The test of any formula of reality is not necessarily its consistency but its being proven in practice, in test after test. Science is enriched by the continual contradiction between models and theory on the one hand and reality on the other. Models only mime reality. We learn about the world *in* the world, whether in the concert hall, a chemist's laboratory, the creation of consistent equations, or in a refugee camp.

The contradictions in physics, music, and a society stuck in long-obsolete views and methods can be resolved only by actual experience, in accepting and acting on the true reality of cause and effect embedded in a process of dynamic change. The alternative is the present Newtonian world in deep distress, powered by a crazily out-of-control engine racing toward darkly unpredictable out-comes.

Is it still within the power of public sanity to avoid dire consequences for our world? Can musicians, scientists, and religious, governmental, and nongovernmental organizations slow down the mad, species-cidal race long enough to find rational, peaceful solutions? Can social and economic crises, including the crisis in science, be found in a true science of society? This would necessitate the marriage of cause and effect based on real conditions and the consequent resolution of the contradictions in order to produce a new, third entity by casting off the old ways of frag-mentation and nearsightedness. We are living in a time when the string section plays its own tune while the woodwinds, brass, and percussion sections each play entirely different compositions, all in unrelated keys. Can this chaos be resolved? If so, where do we begin?

The "Real" Versus the Quantum World

We have seen what the quantitative, objective world of Newtonian science has given us. But we have yet to quantify feelings, literature, sociology, relationships, and, of course, music. How do we quantify what it is to be a human being? How many dollars is a person worth? Can we objectify passion, imagination, eloquence, and the spirit?

The quantum world provides a greater flexibility in approach; the observer now enters the picture. And why not? Even Einstein's relativity in the macroworld takes into account the role of the observer in the measurement of space, time, and matter. Heisenberg's uncertainty principle in the microworld tells us that an observer affects an observation. Whether a particle acts like a wave or a particle at any given moment depends on the observer! This astounding conundrum has been proven over and over again in experiment after experiment. One is tempted to ask: Does an individual voice or vote make waves! Does your vote make a difference? Or your letter to Congress? This writer says "yes."

The question remains as to whether quantum theory really does obliterate the distinction between the observer and the phenomenon, for it tells us only the results of specific experiments. To some degree the same is true for relativity since it measures space, time, and matter with a different "frame of reference." So what is the effect of the individual consciousness or action on the earth, or the cosmos?

Hindu philosophy tells us that all is consciousness. Einstein, like most scientists, puts his faith in imagination, intuition, even dreams. Friedrich Kekule's dream of a snake biting its own tail while whirling in a circle gave him the same insight that led to the discovery of the ring-like aspect of the benzene molecule (leading to much of organic chemistry), which was a forerunner of string theory. Physicist Roger Penrose agreed with Einstein that the functioning of the mind required much more than the laws of quantum theory. The contradictions and controversy continue to swarm all about us. In physics, the opposition between Einstein's relativity theories and quantum mechanics has played a positive role, stimulating scientists to make further discoveries.

Music too, as defined most broadly in these pages, can play an important role in resolving the scientific differences as well as conflicts in our lives. Some scientists keep music on all day while they work. Music can achieve a synthesis of science and society, raising both to new levels of insight and practice.

61

What Is Being Done?

Today, young people get their social studies from
the music they listen to. Students in schools and
college campuses all over the world, in the most unlikely
places, are making their voices heard about war and
peace, freedom and justice. Young people in Ireland
pressed Sinn Fein, the largest nationalist party, to drop
its abstentionist policy. They helped organize the community in urban slums and border counties that led to
the electoral breakthrough.

Women in Muslim countries are, for the first time,
demanding their human rights. In Israel and the Diaspora, a nascent peace movement holds some hope for
change in spite of the thunderous and unrelenting public relations campaign supported by the media, wealthy
foundations, and Jewish and non-Jewish irredentists,
whose policies only spawn an unending cycle of suffering and bloodshed to Jew and Arab alike.

Recent polls in the United States find that leading foreign policy organizations' websites are getting several million hits a month expressing strong support in favor of universal health care, against war, and in favor of greater international cooperation. "People are still searching for answers to some very big questions," said the president of the Chicago Council on Foreign Relations: "In the last four or five months, we have seen a sustained level of interest when the discussion is framed very broadly and we have a speaker whom people recognize." The websites of other foreign affairs organizations are also attracting unprecedented hits. Young and old are taking action. A prematurely wise few even have a long-term vision. College and high school students are linking their domestic problems to a foreign policy thrusting them into war. UN NGOs (non-governmental organizations) report growing worldwide awareness of global affairs.

When I was chairing one of these NGOs, the United Nations Department of Public Information Communications Coordination Committee (CCC), a young teenager (perhaps seventeen or eighteen) came to one of my meetings. Always on the lookout for talent, I recognized a high order of intelligence, a dedication to human concerns, and a quiet self-assurance in this youngster, and I immediately welcomed him and nurtured his political talents. Now, some half dozen years later, Ben Quinto and his friends have organized a worldwide youth organization. With more than 180 branches on every continent and still growing, the Global Youth Action Network (GYAN) educates the half of the world's population that is under twenty-five, and activates young people about the issues that are most important to them, "adding your voice to the Global Youth Agenda for your country, how to be an official Youth Delegate to the United Nations, or debate issues with other young activists online." They counsel youngsters on starting a project that improves their community and offer program models to help empower, organize, and give a voice to young people to influence policies and decision-making.

Crowded with volunteers, their busy main office in New York is never without music, *their music.*

62

Science Courts Music's Vibes

In the sixties and seventies there was a word widely employed—"vibrations" or "vibes." He or she has good vibes, or bad vibes, happy vibes or sad vibes, negative or positive vibes. A roomful of people, a city or even a political campaign had its own set of vibrations. Freely translated, this word described a kind of felt environment and atmosphere, emanations of a group spirit. It described in layman's or musical language what scientists measure as wave frequencies.

It is no accident that the word vibrations had achieved such wide currency. It still has echoes that are interwoven with life as well as music (compare "vibrant"—vigorous with life, resonant; and "vibrato"—achieving warmth and expression on a string instrument by vibrating a finger or by vibrating the breath on a wind instrument). Colors and artworks are said to have certain vibrations, even to the point where some people are said to be able to determine color by touch when blindfolded.

When musicians use the word, they subconsciously imply the fundamental physics of sound, the disturbance of a body's equilibrium, which, by creating waves and disturbing the air particles, transmits sounds to the ear. The pulsations of strings and lips on wood or metal, which find response in the membranes of the ear, are what we call sound. These sounds can be termed music when they approach certain norms of configuration and organization. But these norms change from person to person and from time to time. One person's beautiful music might be considered ugly or boring or cacophonous to another.

Vibrations are measurable on a calibrated scale; electroencephalographs measure brain waves, with divisions according to frequencies into alpha, beta, theta, and delta. Studies of sleep and mind control show that differing levels of consciousness, of mood and creativity, accompany the different frequencies or rhythms. If that sounds like a description of music, it is no accident. The body produces and reacts to the rhythm of life, to the cycles of the moon and the tides of the sea, and to sounds. The effects of the chanting of mantras, the universal *om,* and of meditation have become widely known. Some Indian gurus claim that chanting can change the world. "Harmony," inner and outer, has become tangible. In physics, various synonyms for harmony are in constant use, mostly in terms of balance, symmetry, and, of course, waves. Wave theory must be employed, directly or indirectly, in everything from particles to the cosmos, baby-sized violin strings to grandfather double-bass instruments. Music and science are kissin' cousins.

People Music and People Science

The concept of "field" relates to the tiniest particle, to the entire universe, and to all that's in between, including you and me. We've seen how the magnetic field of a magnet attracts or pulls substances like iron filings. Fields are the areas around all bodies, whether organic or inorganic. The gravity field of the earth keeps the moon from racing off into outer space. This field is similarly beneficial for us humans, though it's a bit inconvenient since over the years it makes our skin sag into jowls and our bodies grow shorter. Closely allied to the word "field" are concepts such as context, surroundings, climate, environment, vibrations, and energy. To cover or describe all of these, try on the concept "music field."

Blowin' in the Wind

Vibrations, based on wave frequencies, lead to a reevaluation of all previous definitions of music. Newly defined, music seeps through everything. As society goes through changes, there are concomitant changes in music and in science. All of us breathe and exchange breath with each other. Ask any nonsmoker in a smoke-filled bar. As we take in oxygen's life force from trees and plants, we breathe out carbon dioxide, which the trees and plants use. There's a constant exchange between all animals and all plants, a give and take. The ecology of the earth has been a magnificently designed interchange of needs. All life is interdependent. Nothing can go it alone.

We humans are the musical instruments of the earth. Our bones, nerves, flesh, and blood are vastly more responsive than the wood, strings, and brass in an orchestra. Our brains and sensory systems are hugely more complex and capable than any computer could ever be. Meanwhile, our bodies, minds, and spirits are producing the waves that could conceivably contribute to a new and better society as well as new art forms. Composers and visual artists have begun to create music and art from human and animal biology, from relativity and quantum physics. What better way for governments, corporations, and foundations than to focus their scientific research and bend their resources and their genius to creating a better world? We have the technology, the researchers, and the artists. Unless we make use of them, we will find that our time has run out.

63

Music and Culture

Culture, art, and music have been defined in widely different ways since the beginning of recorded history. Every age found different answers for each of the arts. In the Western world, culture usually has an esoteric connotation, broadly meaning the fine arts, enjoyed by a minority and held in some suspicion by the majority. Some nations include clean streets in their definition. Visitors from many other countries are shocked when they visit the U.S., where people throw cigarettes and litter into the streets and on the beaches. I've been in foreign cities where those who litter the streets are called uncultured. By that measure or in the fact that a large percentage of the population enjoys classical music and the arts, Germany rightly considers itself highly cultured. Ironically, this national predisposition asserted itself even during the Nazi regime. Though one of Hitler's top aides is reported to have said, "When I hear the word culture I reach for my gun," concentration-camp officers gave musician prisoners special benefits so they could entertain the officers and S.S. troops. But despite a delay in their execution, these musicians were thrown in the same gas ovens as the other Jews.

Culture in
Twentieth-Century China

In China, culture came to include everything through which the people expressed themselves. With a population of well over a billion people who speak many different languages and represent many different cultures, nothing in China is simple, including the political, social, and cultural currents. The contention between opposing positions on lines of socialist development resulted in the Great Leap Forward in 1958 and the Great Proletarian Cultural Revolution launched in 1966.

In 1949, six hundred cultural leaders gathered in Peking for the All-China Congress of Writers and Artists. Traditional female impersonators sat with modern poets and novelists. Popular film stars and ballet dancers sat side by side with Chi Pai-shih, an octogenarian classical painter. The Party policy in the arts—"Let a Hundred Flowers Blossom, Let a Hundred Schools of Thought Contend"—was not working as hoped for by Mao Tse-tung. He stressed that "the intellectuals are reluctant to accept Marxism-Leninism" and that "the proletariat must build up its own army of intellectuals, just as the bourgeoisie does." Before the start of the Cultural Revolution, Liu Shaoqi, then president of the People's Republic of China, believing that socialism had won in China, espoused private plots rather than collectives, and "pragmatism" to develop the productive forces as quickly as possible. This led to clashes between Mao, who thought that socialism was threatened, and leading officials of the Peking Party Committee. Others were drawn in. Mao took action, calling on the people, especially the youth, to play a leading role to "bombard the headquarters."

The response was more than he bargained for, and a nationwide conflagration resulted. The Red Guards—a youth organization—and other revolutionary committees were formed, the educational system was radically changed to help narrow the schism between industrial labor and work in the country. Artists, professors, and musicians were sent to the countryside to "get dirt under their fingernails." The fervor of youth got out of control as students and young people roamed the streets, smashing musical instruments and everything they considered intellectual or bourgeois. Books were thrown into bonfires. Only workers and farmers were venerated.

Not many years after the Cultural Revolution, I was invited to conduct the Central Philharmonic in Beijing. The musicians were either elderly people who had not played for ten years or young, "green" players. The in-between ages were untrained, since most music schools and conservatories had been closed. Fingers and muscles were out of shape. Some instruments had been smashed. At the first rehearsal, my heart sank. Assuming that most professional orchestras had at least a passing knowledge of well-known repertoire, I had sent my program in advance, which included Hector Berlioz's *Symphonie Fantastique,* a fairly difficult work requiring a large orchestra. On arrival in Beijing I learned, to my dismay, that the musicians had never played or heard this work, so I had to start from scratch, patiently teaching and training each section and player, from the first bar onward. Through sheer determination on all our parts the concert was a huge success. The minister of culture said that I had "shaken Beijing." But it was the sheer guts of the musicians who did the shaking. I was declared "principal guest conductor for life."

My brief immersions in China and in their cultural life took me to new levels of realization about the power of the word "culture" and the many permutations and varied definitions of that word. Composers through the ages have drawn on folk and popular music or "borrowed" from other countries and composers. Cross-fertilization in the arts is now so normal—in music, visual arts, dance, and film—that when we encounter "purity" of style, we find it refreshing.

64

Language Power

"Before Jove reigned no busy husbandman
Subdued the ground; there was no usage then
Of landmarks, lines and severance of the fields,
All goods were common, and the liberal earth
Gave every gift unused. 'Twas Jove
Concealed the seeds of fire, stopped the flow
Of streaming rills that once ran red with wine.
Then iron in hot forge
Took temper and the chill-edged saw was made.
For driven wedges first were used to cleave
The yielding grain of wood. Then later times
Brought forth of other arts the varied skill.
Work conquered all, relentless, obstinate,
While poverty and hardship urged it on."

     —Virgil on the onset of class society

Chuang-tzu (c. 290 B.C.E.) made a remarkably similar statement:

> "When the Great Virtue lost its togetherness, men's lives were frustrated. When there was a general rush for knowledge, men's covetousness outran their possessions . . . The next thing was to

invent axes and saws, to kill bylaws and statutes set up like carpenters' measuring lines, to disfigure by hammers and gauges. The world seethed with discontent."

Words at our stage of civilization have become suspect. Ever more frequently they are used to deceive, to obfuscate, to lie. In the mouths of some, peace has come to mean war, law and order means official violence, liberty means repression, and gaining credibility means telling lies. Small wonder that successive generations of young people satirize words. "Cool," "crazy" and "bad" mean good; "freaky" and "weird" have desirable connotations; "outrageous" and "awesome" mean especially favorable. The one language that resists misrepresentation is music! What you hear is what you get, for better or for worse.

From biblical times onward, words and music have served one master or another: the "awful sound of the trumpet" (ram's horn) called people to prayer or war; the birdsong called a mate; spirituals signaled the slaves to escape routes on the Underground Railroad. But nobody ever called a middle C a lie or went to war because their views on a triad differed. Music, because it is honest, has always been the Answer to many questioning youngsters. Music is believable.

The Bach B-minor Mass regaled the church audience for the greater glory of God. The tortured introspection of the nineteenth-century Romantics—Schubert, Schumann, Brahms—expressed unrequited or cosmic (in the case of Beethoven) love. Music probes feelings, emotions, even ideas. The last great Romantic, Wagner, went on endlessly, from modulation to modulation, idealizing a heroic master race (at least that's how Hitler interpreted it). The "talking drum" of Africa conveyed a basic language, signaling commonly understood information at great distances.

When I conducted the first ever symphony concert in New York's venerable Apollo Theater in Harlem, where many of the great black artists got their first chance to perform, the response was equally surprising. It was the time of the black riots, and though I had many friends from my previous project, the Harlem Music Project, which trained string players and was successful in introducing black instrumentalists into major symphony orchestras, this background was to no avail. My life was threatened. How can a white symphony conductor dare to perform in their turf! No matter that my orchestra had an African-American woman timpani player and the program featured a black jazz quartet as well as a symphony orchestra. Nervous but undaunted, we arranged for the cooperation of Malcolm X's widow, who arrived all dressed in white, with twelve cleanly starched children in tow. They sat in the front row while Malcolm's guards, dressed to the nines with black suits and immaculate white collars, served as ushers. To cap it all, Duke Ellington's sister Ruth, on my right side, and the great actor Ossie Davis, on my left, accompanied me from the wings to the podium. Both Ossie and Ruth addressed the audience, entreating their good behavior and cooperation for their friend the conductor.

The concert went smoothly, and much to our surprise, the serious symphonic portion of the program elicited greater applause than did the rock and jazz portions. Could that mean that there is a huge, untapped audience out there, somewhere, for classical music? Could this unusual experience have an even deeper significance?

The Power of Music's Truth

Unlike words, music cannot obfuscate or lie. Sounds can be trusted. Whether they are tender, violent, sensuous, mocking, satirical, diffident, affirmative, nostalgic, yearning, blue, or joyous. Young people talk of good music or bad music. The criterion for audiences young and old, the music they react to, though they wouldn't express it in words—comes down to "Is it believable? Is the artist doing his thing?" Is he or she expressing something that they recognize in their own lives, or is it just the acrobatics of virtuosity? What kind of person is the performer? John Lennon and Sting gained particular credibility because of their modest backgrounds and because they stood for a principle; in the case of Lennon, it was peace.

If serious musicians are to gain the ears of the new generation or avoid extinction altogether, they must take a hard look into themselves, their art, and the times. The greatest musicians and scientists were those who were constantly critical of their work, their society, and themselves. They had a vision well beyond their personal world.

65

The Unanswered Question

One of Leonard Bernstein's obsessions was the language of music and its relation to spoken language. He explored the subject in a fascinating book, *The Unanswered Question.* The title comes directly from a 1908 composition by New England composer Charles Ives. Bernstein's book, published in 1976, is based on six talks he gave at Harvard in 1973 as part of the Charles Eliot Norton Lectures. The great linguist at Massachusetts Institute of Technology, Professor Noam Chomsky, further sparked Bernstein's interest. Linguists and many others are still debating, after more than forty years, Chomsky's landmark theories. His views on human language are being taught or discussed wherever language is studied. The core of his theory is that human beings are unique in their possession of brains equipped for generating grammar. In other words, some language skills are innate. He presents considerable evidence that very young children are able to construct grammar and a whole language, "working at it together or, more likely, playing at it together." The stories about the innate response of babies and small children to music are legion, and most children show some kind of musical talent under the right conditions.

We know about the ease children have in learning languages. Sadly, we lose that ability as age comes upon us. I've been trying to learn Spanish for some years, slogging through tapes and books with too little to show for it. Moreover, as a second-generation child of immigrant parents, I rejected the opportunity to learn my parents' Romanian. And though I remember a certain amount of Yiddish, I was ashamed of that language, too. As the lone Jew in my class, I wanted to be "100-percent American." What a loss! Now I relish Yiddishisms, as does my Catholic wife.

Among the many who were affected by Chomsky's theses was biologist Lewis Thomas (*Lives of a Cell, The Medusa and the Snail, Late Night Thoughts on Listening to Mahler's Ninth Symphony*). Thomas points out: "If it were not for children and their special gift, we might all be speaking Indo-European or Hittite, but here we all are, speaking several thousand different languages and dialects, most of which would be incomprehensible to the human beings on earth just a few centuries back."

One wonders if Chomsky's virtuosity on the question of language was somehow related to his uncompromising left-wing writings. (Surprisingly, he considers his linguistics work separate from his political views.) He defends Palestinians, poor people, and those who are unjustly accused or in prison. Bernstein was probably attracted to Chomsky for these qualities as well as for his incontestably distinguished scientific credentials.

Conductor Bernstein tells his audience of Harvard students, "Perhaps the principal thing I absorbed from Harvard in general, was a sense of interdisciplinary values—that the best way to 'know' a thing is in the context of another discipline. The lectures were given in that spirit of cross-disciplines; which is one reason why I speak of music along with quixotic forays into physics."

Bernstein's wide-ranging mind covered many subjects. He steeped himself in oriental philosophy as well as its music; his knowledge of music was encyclopedic. He became a liberal force for civil rights and a peaceful world, and if it were not for his prominence, he would have paid a stiff price during the McCarthy witch-hunting period and during the Un-American Activities Committee hearings. As it is, he was deeply hurt when Tom Wolfe satirized him for his flirtation with the black radical movement. Lenny was a sensitive soul and never got over the snide attacks from a bestselling author.

He admits to being "haunted" by the notion of "a worldwide, inborn musical grammar." In explaining the reason for the title of his lectures, Bernstein conjectured that "Ives had a metaphysical question in mind; I have always felt that he was asking—'whither music?' We must first ask whence music? What music? Whose music?"

Bernstein addresses these questions in his splendid book, beginning with a vivid recollection of a recording of "Piano Variations" by his friend, Aaron Copland. On analyzing the composition, he made the "startling discovery" that the first four notes, the "germ" of the work, are also the subjects of such works as Bach's C-sharp-minor Fugue from the *Well-Tempered Clavier,* Stravinsky's Octet, and Ravel's *Spanish Rhapsody,* and is in the Hindu music used by the Uday Shankar Dance Company and in the music of other non-Eurocentric sources. Like Einstein's simple equation of energy with mass and the speed of light, it attests to the basic simplicity of music and art. Upon this foundation is constructed the variety, towering complexity, and power of these arts.

66

Art Power

Science had its heyday during Sputnik and then gradually faded until the eighties, when string theory came to the fore. Like Sputnik, the arts explosion in the sixties and seventies burst on the scene with such incredible force and in such disguised form that even those of us who have spent our lives in the arts were slow to recognize its implications for our lives and for the human environment as a whole. Intuitively, artists were breaking down boundaries and traditional forms, moving into streets, canyons, and subways. The business world, recognizing a good thing (more so than the artists themselves), started a trend of paying ever larger fees to buy up the creators, creations, and practitioners of the arts and sports to sell their products. Their investment paid off handsomely. Soon, even presidential elections and campaigns began to be aptly termed "showbiz," and brought historic profits to the corporate and political investors.

If showbiz can be termed an art, then one can suppose that this recognition of the power of the arts was nothing new. When the premier of Japan, head of one of the great capitalist nations of the world, visited Mao Tse-tung, poet and head of one of the great communist countries, he wrote a poem in commemoration. (Try to imagine a U.S. president doing this!) The Bolshoi Ballet danced their way through the "iron curtain" between the U.S. and Russia. Prior to impresario Sol Hurok bringing the Bolshoi to the U.S., we did not "recognize" the USSR. There was a huge empty space on maps where the USSR should have been. The art of Ping-Pong opened the locked gates between the U.S. and China. The power of chanting (a kind of music) from the Eastern philosophies, imported into the West by Yehudi Menuhin and the Beatles, affected a large swath of musicians and young people, opening their eyes to a larger world.

The arts have such power that it staggers the imagination. From biblical times, artistic "symbol" and the musician's cymbal were employed to equal effect, and for thousands of years the Church subsidized the arts for the greater glory of God, bringing even greater glory and profit to the Church. Hundreds of millions of people watch the theater of TV, a moonshot "show," an assassination "show," a political convention "show," and an intifada "show." The World Trade Center tragedy and the Iraq war provided unending grist for the television mill, with real characters being killed. Our homes, walls, appliances, and gadgets are all created, designed, packaged, promoted, and sold through the arts. Imagine if all the artists (fine, performing, commercial, industrial) were to go on strike. Our economy would come to a screeching halt. On the other hand, imagine what the same technology, in a rational, scientific social system, could do to help free the peoples of the world from want, repression, and misery.

Once, on tour with the Eger Players, a chamber group consisting of piano, violin, cello, and French horn, we were invited by Judy Stone, *San Francisco Chronicle* cultural editor, and Warden Douglas Rigg to perform at the Minnesota State Penitentiary and at San Quentin in California.

Rigg, a tough but fair warden, had introduced training and social pro-grams, winning the respect of the prisoners, many of whom went on to good jobs. Both prisons housed "lifers," those who had committed mur-der. They were a rapt, enthusiastic audience for our largely classical mu-sic program. At a post-concert reception, we were warmly greeted by individuals, apparently sensitive and educated men, many of whom, in a moment of desperation, had murdered their wives. Mostly Catholic, they could see no way out, since their religion had no space for divorce. Our music obviously fed their souls.

67

The Physics of Music:
The Music of Physics

"Critics can't even make music by rubbing their back legs together." —MEL BROOKS

"In opera, there is always too much singing."
—CLAUDE DEBUSSY

The quantum revolution, radical as it was in its mystical and spiritual aspects, had roots extending deep into history. The great musicians, whether or not they put it into so many words, embraced the spiritual and mystical view of the world. Many were touched by Eastern philosophy, either in their reading or in visits to India, China, Japan, Greece, and Egypt. Those in the scientific world who struggled with the ancient concepts included Einstein, Bohr, Schrödinger, Eddington, and Heisenberg, to mention a few.

Quantum science captured the popular imagination after publication of books such as *Music of the Spheres* by Guy Murchie, *Zen and the Art of Motorcycle Maintenance* by Robert M. Pirsig, and *The Tao of Physics* by physicist Fritjof Capra. Lawrence LeShan, author of *The Medium, the Mystic, and the Physicist,* once shared a day with me, explaining his then unusual science/music relationship. His book aroused passions and controversy that only increased once quantum mechanics reached the public's imagination.

My first small foray into the debate started in the fifties when I was on the faculty of the Aspen Festival in Colorado. I gave a lecture/workshop for singers and wind players called "Breathing for Life and Artistry," in which I attempted to tie together spirit, esprit, the spiritual, breathing, and quantum mechanics, all with music! Luckily, my audience included no physicists, though in a later lecture in New York I had plenty of room for intimidation since my audience included Gregory Bateson (Margaret Mead's famous husband), who genially stated my thoughts in more scientific language.

As a fundamentally skeptical person, I never ceased to straddle the two camps, the scientific and the spiritual. At age fourteen, I had what seemed then to be a clear vision of relativity (and even a crude quantum mechanics, though I wouldn't know the term until years later). "Of course," I thought, "every breath, every musical wave affected every other breath, wave, and particle throughout the universe." It was abundantly clear to me that we do indeed affect the world, just as the world affects us, in complex, dynamic, and ever changing patterns. That poor butterfly, flapping its wings, sure has some responsibility!

Change

The relativity of space-time and the interconnection of all things, at least on the quantum level, have permeated—if foggily—the consciousness of the lay public. Since understanding can lead to action, the power elite senses danger to its prerogatives. It responds by tightening its control over information, whether that information is "the news," the media, the arts, or as much of the freewheeling entertainment world as they can control.

Held for centuries under the thumb of the economic, political, and social elites, the peoples of the world may be approaching a level of indignation where, with "nothing to lose but their chains" (Marx), the "hundred monkeys" syndrome may set in. This is the point where enough people will collectively shout, as in the film *Network,* "I'm mad as hell, and I'm not going to take it anymore." The "hundred monkeys" syndrome holds that when enough monkeys gradually follow an initiator into a behavior change, the entire species will follow suit, suddenly rather than slowly. Is the equation of truth to power in the monkey's great-great-grandchildren similarly shifting? Seems that it is gradually becoming apparent that the old, Newtonian ways of thinking in the framework of separation and fragmentation may require a bit of tinkering.

Security

Historically, the *means* to wage war and preparations for "national defense" have usually meant that there *will* be a war. In the name of "security against terrorism," real security of the individual as well as the most basic human rights are violated. Science has given us technologies that permit war against others and our own people on a nightmare scale. Scientists can be found who sell their skills to the development of killing machinery. Others sell their discoveries to pharmaceutical corporations that depend on research scientists to increase profits and fatten the bottom line of the company stock. Whether the drugs and chemicals are beneficial or poison our food, environment, and bodies is considered to be a bothersome irrelevancy, best handled by their legal department.

Loving Obsolescence

It bears repeating that all former explanations of Newtonian and Cartesian reality became obsolete when general relativity, quantum mechanics, and superstring theory arrived on the world scene. Paradoxically, the same paradigms of Newton and Descartes, including all nuances of scientific method and classical mechanics, laid the foundation for this world of evolving technology and discovery. Before these gentlemen, rationalism and technological change were resisted, as were the discoveries of Copernicus and Galileo. The basis of modern astronomy was set in the Copernican system, which posited that the planets revolve around the sun and that the turning of the earth on its axis accounts for the apparent rising and setting of the stars. Galileo, we remember, was condemned for heresy by the Inquisition in the seventeenth century.

It is hard to believe that similar ignorance is appallingly widespread today. The resistance to the full implications of the twentieth-century scientific revolution goes beyond the movers and shakers of policy and the media prostitute dealers in information. The blinders cover the eyes of educators and scientists as well. Whether knowingly or not, we remain in a Newtonian mode, while the world's business and politics are conducted in crude adaptations of feudal thinking. This obsolete approach poses a distinct cultural and scientific barrier to acceptance of a more dynamic, comprehensive, and sustaining paradigm.

To make it worse, fears of war, "terror," pestilence, famine, and death remain a shadow on the light of change. The falsehood is in accepting this "prison" as inevitable. Yet these notions are based on an economy of scarcity in the midst of plenty. For the first time in human history, we have the technology and the resources to feed and care for our earth and all its people.

Way back, though people had fire and a primitive level of technology, they were always hungry, having to share in order to survive. Through the millennia, they coped with nature by ever more advanced uses of technology. The level of this technology varied in parallel with advancing social relations. The domestication of plants and animals, for example, only eight thousand years ago, gave a modicum of assurance of a stable food supply. In turn, this new method of "making a living" radically changed social relations. Once people had enough, followed by a surplus, they had stuff to trade, eventually including property. As technology advanced and the population increased, we passed through the feudal epoch to the slave-as-property epoch. These periods were superseded by market capitalism at the national, state, and eventually international level. Through it all, technology was the leitmotif of social change and structure.

One day, Henry Ford proudly showed union leader Walter Reuther a factory that needed only one worker. Reuther responded, "Who then will buy your cars?"

Thus does the advance of technology produce an anomaly. The capitalist economy lives only by selling goods, depending on people with enough means to buy these goods. More than eighty countries are poorer today than a decade ago (UN statistics). Social programs are in rapid decline as a result of war. The production of weapons tends to give a temporary uplift to the economy, but political and cultural life degenerates. Science and the arts suffer. Music, science, and society are in constant concatenation with social turbidity. The linkage is inevitable. Only by sacrificing for peace will truth and peace become a fixture of the human condition.

68

A Revolution in Physics and Music

U nder the heading "Atomic Scales: Striking Notes of Progress on the World's Tiniest Guitar" (*The New York Times,* November 9, 2003), science writer George Johnson followed up on a previous discovery of a black hole, 250 light-years away, "humming a bass note 57 octaves below middle C." This time, physicists at Cornell University used a laser beam to pluck the invisibly tiny silicon guitar. Each string is about 50 nanometers (billionths of a meter), and when plucked, the sound is 40 million cycles per second, or 17 octaves above what any human ear would take for music (which is 20,000 cycles per second at most). Dogs and mosquitoes beware!

Though this tiny "guitar" is the stuff of subatomic quantum theory, it apparently has lots of practical potential, while it also replenishes the feeding troughs of science fiction. It also confirms once again the theme of this book that the universe is made of music. By this time the reader knows how broadly music is defined in these pages and how its power has been manifested through the ages. But to know either science or music in their fullest perspective, one has to enlarge the frame of reference to include social science, economics, philosophy, politics, psychology, and many other disciplines. All of these are affected by and in return affect societal change. In order to fully understand these changes, it is necessary that they be examined in their full social dynamics and interconnections.

Cross-disciplinary recognition has become common. Professor Jeffrey Sachs, considered one of the world's foremost economists, having advised reforms in Poland and engineered radical changes in Russia (where I met him and enjoyed a chat about chaos theory among other things!), was formerly professor of economics at Harvard and is now director of the Earth Institute at Columbia University. His basic field of economics is now merged with such diverse subjects as the environment, women, and health. He is UN adviser on HIV/AIDS and international development, and he heads a huge effort to help poor countries deal with disease. Dr. Robert Mellins, internationally renowned asthma specialist for whom Columbia Presbyterian Hospital named a newly built wing for children, is similarly involved in other disciplines, including music. The connections between seemingly disparate subjects are becoming increasingly common. Condoleezza Rice was a concert pianist, as was an advisor (now deceased) to the Sharon government. Scientists are notorious music lovers, and some are quite accomplished.

Music's pervasive influence is felt in every corner of the world, just as the dramatic discoveries over the last century make physics equally omnipresent. The long-term influence of the sciences on society has been well documented, from those brave souls centuries ago who overthrew superstition and the constraints of the Church in the Middle Ages, to the twentieth century's quantum changes and their unavoidable impact. One need only mention Einstein's and Bohr's public complaints about the dangers of the nuclear bomb. Other scientists have followed suit, though one must admit that some, for reasons of family, misplaced patriotism, or personal and career reasons, lent their genius to creating ever more devastating bombs, gigantic war machines, toxic chemicals, human cloning, and genetically modified everything.

Though history is loaded with other wars, plagues, epidemics, and natural disasters, we have our own seven plagues, more threatening than any previous. Choose your favorites: nuclear weapons and waste, AIDS, biological terrorism, global warming, impoverishment of half the people of the world, "local" wars, global environmental degradation, the growing incidence of cancer, state and individual violence. How shocking it is to read about children killing children and their teachers, parents killing their children, and children killing their parents!

Paradoxically, the global economy, widespread environmental disasters, wars, and an expanding conflict between a corporate/political administration and the masses of frightened and suffering peoples worldwide, may provide the tools for change. Though individuals feel isolated, divided, and impotent in the face of overwhelming state and corporate power, hope is in the wings, pushing its way onto the stage, rustling the carefully concocted scripts of spinmeisters such as presidential speechwriters and advisers Karl Rove, Paul Wolfowitz, Richard Perle, Donald Rumsfeld, Condoleezza Rice, and Vice President Cheney.

At the same time, advancing technology works both ways, serving individuals, NGOs, and people's organizations as well as repressive forces. The Internet not only sells books and bangles but also carries messages by peoples' organizations, messages that tell truths which have been twisted and mangled or smothered in the mainstream media. Computers are still owned by a relatively small portion of the world's people, but the messages are getting out. The current wars (Iraq and other targets) are less of an easy sell in the face of the peoples' media—via computer. There we can hear words of caution by members of the international community as well as our own citizens, including scientists and musicians.

Americans are becoming educated as they see their life savings disappear while the rich slink off with billions of our tax dollars. Corporate CEOs get away with a tap on the wrist for cheating their investors and employees while the less fortunate and black males are piled into prisons for minor infractions. Clearly, political leaders, with their armies, police, and weapons, have been unable to prevent terror, drug abuse, and pollution, or to cow people in tiny nations. The governing stages need new actors and directors. The old script is simply unworkable. Does it need to be scrapped in its entirety? Then what?

How about an "I Have a Dream" vision?

A (New) World Symphony

A new world symphony would integrate all the instruments and "sections," providing a new, heavenly music, a body of knowledge about nature and the arts in their interconnectedness—nature and art as dialectic. State forms and class forms would disintegrate or change to allow the healing of the fatal gulf between theory and practice that these forms generated. This is no longer a utopian dream but a necessity if we are to solve current crises and allow the emergence of a new worldview as the condition for the shattering of the old. A new morality and ethics based on the real conditions of living rather than on abstract philosophies. Do people everywhere have enough food, clean drinking water, shelter, educational opportunities, health care? These questions would be the elements of a new world symphony.

The process toward change won't be easy. Status-quoers will fight to the death with many powerful weapons at their disposal. They recruited scientists to provide the technology for their wars and internal repression. Music and the arts have also been heavily co-opted for the media and in the fundamental Christian world, which created their own religiously based, "family"-oriented infotainment, all subsidized with endless cash from conservative foundations. Tune in to the religious stations to get a whiff of the banalities, numbing the minds of a consuming public.

Musical talent is in plentiful supply to write clever jingles and to discover ingenious new ways to sell ideas as well as soap, chemicals, tobacco, and gadgets. Most, however, are not inclined to sell themselves and their art as they struggle to express their social views in "subversive" song. Traditionally, the arts and sciences provide guideposts toward a new and more humane world, though artists are constantly tempted to submit to their legitimate need for financial security. Digitized versions of songs that can be distributed online have challenged the giant record industry, which is furiously trying to find ways to control this runaway people's technology whereby consumers may benefit by the lowering of the price for CDs and DVDs.

Change

Students, always in the vanguard of demand for change, complain of lonely, individualized study and compartmentalization. No wonder they welcome cheap, available music. They sit before the computer screen while in many cases they are able to get degrees without ever going to a class with other human beings. Since universities are often dependent on government, foundations, and the defense industry, the teaching of critical thinking is replaced by "facts" and grades.

The backlash to this unprecedented repression is the gradual education *by experience* of young people who are forced to think critically and contextually, to eschew oversimplifications, to better understand the roots of controversy and to take into consideration the desperate needs of peoples. The young are beginning to gain the ability to take appropriate action. The previously mentioned Global Youth Action Network (GYAN) is a growing organization that facilitates collaboration among youth and youth organizations that are "committed to uniting their efforts to improve our world." With hundreds of organizational members and active partnerships in almost two hundred countries and territories, the network is increasing its outreach and promoting collaboration to improve youth engagement and participation in decision-making. GYAN and TakingITGlobal, a sister organization, have partnered to create a dynamic and resourceful online community.

A new consciousness is coursing through the world, a consciousness that is beginning to realize the necessity for the world's suffering people to unite and cast off the forms and practices that threaten their future and the future of their children. This Achilles' heel of the imperial power—a unity of peoples—could create a new economy, a new science, a new music, a new world. Young people in particular need guideposts toward that future. A lesson or two from the past is a good start.

69

Civilization and the
Good Society

In previous centuries, civilizations emerged as solutions to a relative scarcity between what we could produce and our material needs. In some part, civilizations arose so that ways of life, which fulfilled the needs of privileged minorities who had achieved hegemony in social, economic, and political life, could be rationalized. But we no longer live in a low state of technology. We can now satisfy all the needs of all human beings on the planet. Either we use these mega-technologies to create a global life of abundance or we will become victims of our own technology.

In modern times, the field is littered with philosophers, economists, and even a few government leaders who took a crack at creating a map to describe "the good society": Marx, Lenin, presidents Wilson and Roosevelt, Parson Malthus, Henry George, Adam Smith, John Stuart

Mill, Mao-tse Tung, Thorstein Veblen, John Maynard Keynes, and a slew of utopians. All have fallen by the wayside. Einstein in 1949 wrote "A Plan for Peace," and in 1956 "World Peace Through World Law." Then, through personal visits and letters, he attempted to engage the political elite and moral authority figures throughout the globe in the notion that it was possible to transform the present system of militaristic, unilateral decision-making on the part of nation-states into some world federalist system in which war would be eliminated and there would be global institutions to handle claims, grievances, conflicts, and international norms of behavior. The United Nations was created for a similar purpose.

All of these noble efforts shared one failing; they operated within the paradigm of the obsolete nation-state system. This may prove to be a mixed blessing since it is not only the oppressed peoples who now have the potential to rise up against their oppressors, but resentful nations may themselves become temporary allies in the struggle to get out from under the boot of American imperialism.

Fascism Revisited

In my teens I remember reading about Berlin's Reichstag fire and the rumors that it was planned by the Nazis as their excuse for the takeover of the German government. As fascism began to show its ugly face, some of Germany's best physicists and musicians—at least those who could get out in time—fled from Nazi Germany. Einstein was the most famous of these. American musicians kept hearing about German-Jewish musicians who were humiliated, lost their jobs, and were thrown into cattle cars for the journey to concentration camps and fuel ovens. There were claims and counterclaims about the great German conductor Wilhelm Furtwängler. Was he a down-the-line Nazi, or did he protect and save his many Jewish musicians? The argument was never resolved, though both positions had an element of truth. When he was invited to the U.S. after the war to guest conduct, the protests were so vociferous that he never did conduct here.

An exquisitely crafted film, *Taking Sides* (2001), takes place in Berlin shortly after the war and focuses on an American army major whose job is to ferret out highly placed Nazis to be put on trial. His particular assignment is to "get" Wilhelm Furtwängler and to expose him as a Nazi who was cozy with Hitler and other Nazi officials. The major, played in white-heat anger by Harvey Keitel, is determined to break down the austere musician, who is highly revered by every German as "the greatest conductor in the world." Two other principal, keenly nuanced characters, the major's German secretary and his American-Jewish lieutenant, music lovers both, are awestruck by the presence of the conductor and become visibly uncomfortable as the major loudly denounces the conductor to his face as "a piece of shit" in the series of interrogations.

Furtwängler comes off as a passionate musician and a consummate ivory tower artist before being broken down by the major's onslaught. He had always kept art and politics separate. But during the course of the savage hearings, he comes to understand that music and politics cannot be separated.

I was deeply moved by the film, for I too revere Furtwängler's music-making, moreover, my own family was not immune to the Holocaust. What particularly struck me was the director's apparent intention to portray the American's raucously brutal tactics as similar to the behavior of the Nazis. The film implies parallels with the current situation in the United States, where music and politics are among the victims of the corporate government.

9/11 has been continually used as an excuse for the government to crank up public fears so they could start a long-planned war against Iraq as the first among other Middle East states, meanwhile robbing the public treasury. The ensuing fiats in the name of "defense against terror" were similar to Hitler's blaming of the Jews for German and the world's problems. Admittedly, we have our own American brand of regression. We do things bigger and better. The peacefully won coup was quickly rubber-stamped with an "election" by cronies on the Supreme Court. Soon to follow was a rejection of the World Court, scrapping of international treaties, a takeover of the judicial system, the expelling of foreigners, putting "troublesome" minorities in jail, and the erosion of civil liberties and Constitutional rights. Typical of the German fascism I remember from the thirties and forties was the ballooning of funds to the corporate/ military and the brandishing of military might against any selected enemy of the moment. Both coups were well-oiled campaigns. But our corporate kings collected booty greater than all the pirates in history combined.

Domestic and foreign economies were skewered as the huge super-structure of credit, loans, and make-believe capitalization, foisted on "developing" country after country, toppled down on the heads of the poor and the disenfranchised. Scarce resources were used to pay interest on the debt while people starved. Many Americans too saw life savings drained as "outsourcing" sent jobs overseas, where workers were glad to work for a small fraction of American wages.

Heal Thyself

Can the free-market system heal itself? It has done so in previous crises, notably after the Great Depression, when President Roosevelt took heroic measures. The rise of Hitler and World War II with its concomi-tant production of war materials gave capitalism another shot in the arm. The prime rate was fiddled with and other temporary measures were taken during the several recessions since, but economists don't see the gears quite meshing this time.

Having hijacked the treasury's reserve, the political and corporate leaders are blind to the economic and social science that could save even them and their children from environmental and economic catastrophe. Power and greed have closed off the danger/response mechanism that nature has endowed to all animals. The only hope is that the many frag-mented groups of oppressed and exploited peoples (even those who would not consider themselves in these categories) will unite to save themselves, our species, and the very earth itself. Church, student, labor, and civic groups are beginning to combine with ordinary citizens (in-cluding enlightened business groups) to fight for peace, justice, and a sustainable planet. The promoters of single issues are coming to realize the linkage between their and others' issues and that we all share the same stage. Health care, poverty, the environment, injustice, women's issues, and peace are interlocking issues. Though the corporate media has been very clever in obfuscating the real reasons (oil, money, and power) for their war, experience on the ground is breaking through the swarm of lies. The words "peace" and "compassionate conservatism" have been used to cover up their opposites in actual deeds. The misuse of these words provides a convenient fog to cloud the realities of injustice and destruction of human rights. In the Middle East it has driven hundreds of weaponless men, women, and children to destroy themselves after their land was taken over, their homes destroyed, and their brothers and fathers humiliated, imprisoned, and tortured.

But the occupying power destroys itself in the process. Partners in crime become increasingly vulnerable, while innocent Spaniards, American soldiers, Israelis, and Palestinians die. Anti-Americanism and anti-Semitism grow in country after country. State terrorism begets individual terrorism. The snake feeds on its own tail.

The world's most powerful leaders tell us that through war you can obtain peace. This idiocy is promptly picked up and promulgated by their fully owned media. Their corporate friends win huge war contracts. The craven corporate pundits overwhelm the airwaves with moronic blather. It all comes down to a choice: to continue reverting to a world of misery and greed or to struggle for truth and plain sense. Which will we bequeath to our heirs, the insane destruction of all the wonders man has wrought or a universe of music? The choice is on the agenda as never before.

70

Is It Too Late?

Can music and physics point to a solution? Can mechanistic thinking and the resultant action or inaction be transcended? These could be the fundamental questions for humankind at this stage of civilization. Artists and scientists may be in a more favorable position than politicians to work at the answers. Jointly, they have uncommon abilities that, individually and in combination, are more closely suited to finding solutions. They have a more consistent, day-to-day need to apply integrity, imagination, and vision to their practice and performance. They must consider alternatives to what isn't working. They know about seeking order out of chaos. Their everyday work has to do with workable and practical left-brain solutions combined with right-brain spirit and compassion.

Can new actors and directors (read government leaders and corporate heads) save the planet? Though Democratic politicians have a somewhat lesser stake in the corporate takeover and may be a bit saner, one can wonder if "elections" can repair the broken capitalist system. After oft-repeated false predictions, it could be nearing collapse. By the time you read this, there will be either the same old regime in power or a new "lesser evil." The politicians will promise everything until they attain office, all with great public relations fanfare. But, based on previous disappointments, less of the public than ever are so naive as to believe these promises. Moreover, apart from the common commingling of elections and dirty tricks, well-placed cynicism has kept a large portion of the public from voting.

The attempts to save the capitalist system from its mounting cyclical weaknesses goes back almost to its beginnings; each depression was overcome by varied benefits and rescue missions. During "good times," productivity of goods favored a fair percentage of the public, advancing their quality of life, though many were left behind. Farsighted leaders such as President Franklin Delano Roosevelt helped with government programs that gave the system a new lease on life. World War II added a big second wind. Tinkering with the economy kept pushing it along. Saviors emerged just in the nick of time: economists, philosophers, and academics together with the politicians and generals who backed up emergency measures with political and, if necessary, military power. But the free market's ability to pick itself up and survive has been waning over the past century as it runs into unprecedented global complexities and problems. The military, the courts, and all the king's men are being called upon to prop up a dying system.

Adam Smith, the first of the great economists, lived in a world of a social free-for-all, a truly competitive market mechanism. He scoffed at those who hoped to improve society by "doing good." John Stuart Mill pointed out that economics had no ultimate solution, so he introduced moral judgment into the equation. Hence began the notion of "interfering" with the market process, continuing in various forms to this day. In the factories of Mill's time, specialization in a Chaplinesque sense was in its infancy. Smith's *Wealth of the Nations* (1776) gave impetus to both self-interest and regulation of competition. He explained how society could induce its producers of commodities to provide what it wants. He found in the mechanism of the market a self-regulating system for society's orderly provisioning. He taught that one can do what one pleases, but if the market disapproves, the price is economic ruination.

In the late nineteenth and early twentieth centuries, Thorstein Veblen faced the more advanced world of railroad magnates, the U.S. Steel Corporation, and the rapid growth of big business. His infatuation with the machine led to his thinking that a corps of engineers, those who really ran the factories, could run our society better than the owners, who were less interested in production than in profits. He saw his world as savagely run by "absentee ownership," by a leisure class that was predatory and with deep-rooted, aggressive tendencies. He died before the great '29 crash.

Herbert Hoover, basking in the prosperity of the bull market of the twenties, exulted: "We shall soon with the help of God be within sight of the day when poverty will be banished from the nation." He promised "a chicken in every pot." The chickens died in the Great Depression. President Roosevelt then tinkered heavily with the market with his government-sponsored social and employment programs. Many of his valuable constructions, such as Social Security, are under coordinated attack by wily corporate and political leaders, who want to grab for their own "free market" pockets the life savings of working people's old-age insurance. FDR probably learned from the previous Roosevelt, Theodore, who became popular by attacking the oil barons and by introducing legislation to regulate big business. Current leaders don't feel they need even pay lip service to such endeavors.

Hitler and World War II finally rescued the capitalist system. It was a time of full employment, as the nation mobilized for war. It came to be the accepted wisdom that war was good for the economy. Indeed, some great fortunes were amassed—by a few. At the same time, regulations were put in place to set a balance and to protect the public. These regulations were in place until recently, when the Bush administration's corporate cronies aimed to abolish them. In their place are mammoth tax cuts for corporations and upper incomes, ostensibly to promote business investment. This step was touted as a creator of new jobs, the "trickle down" theory. Instead, jobs soon began hemorrhaging. A $200 billion surplus swung to a $170 billion deficit, increasing with military outlays to what could be a $400 billion-plus deficit. But Wall Street, in its infinite wisdom, was not so sure. The stock and bond markets rocked back and forth. Other nations, large and small, faced similar or more severe problems. On the horns of a dilemma, cornered capitalism showed its ugly teeth.

The system's imperial brutality grew like a fungus. While entire communities live in privileged, guarded, and gated prisons, others are desperately trying to survive, by stealing if necessary, to get a slice of the great wealth capitalism generated for the few. Building prisons and packing them with potential "troublemakers" has become a growth industry, a good investment for stock speculators. Whether rich or poor, everybody is frightened and insecure, each within his own, separate prison.

Though the British invention of "divide and conquer," now thriving in its former colony, still manages to sharpen divisions between nations, religions, the haves and have-nots, blacks and whites, young and old, women and men, these divisions may be closer to collapse as consciousness grows that the same problems are shared by all. In a still nascent but defiant cross-border solidarity, the slowly growing realization takes hold: The bare subsistence worker in Mexico, the 27-cents-an-hour Chinese laborer, and even the highly paid American worker have issues in common. There is one common enemy, the global corporations and their giant home Nation. Old-fashioned Marxian categories of proletariat versus industrial mill owner and the setting of ever weakening labor unions against their brothers in other countries is now complicated by globalism, tariffs, and protectionism.

Previously unheard of alliances are slowly emerging out of the realization that solutions may be found only by uniting against those who are tightening the screws of regression and repression. Nature can't be fooled. Her catastrophes don't discriminate between human categories. Global warming takes over the entire planet. Toxic fish and chemical foods invade the gourmet palate as well as the workman's lunch bucket. While the president and his corporate family live in ecologically correct houses with their own water-purification systems and geothermal heating and cooling systems (water at a constant 67 degrees, piped up from 300 feet below ground for heating the swimming pool, cooling the house in summer and heating it in winter), they are rushing pell-mell to roll back and abolish environmental regulations and treaties. Meanwhile, whole populations are dying for lack of clean water, which is only one of their problems.

71

Beginningism and Endism

Until the mid-1980s, a slowly growing cloud of depression had enveloped theoretical physicists. John Horgan's *The End of Science* epitomized this depression. There seemed to be little more to learn about particle physics and the world of science in general. The horizon was bleak, with no sign of new frontiers. Horgan's endism was nothing new. We had Oswald Spengler's massive *Decline of the West*, Francis Fukuyama's *The End of History*, Madison Grant's *Passing of the Great Race*, and Lothrop Stoddard's *The Rising Tide of Color*.

The End of Science cynically purported that all the great fundamental discoveries have already been made. It contended that the paradigms or mechanisms that explicate events are established. Our fears now included technologically associated fears. There was the Y2K fear that worldwide computer failure would shut down transportation, business, and agriculture and would impact the entire economy, from the stock market to livestock, violently shaking the computer-dependent "advanced, industrialized world." This hysteria was compounded by new/old fears.

The death of music was also predicted, though the mourning had begun decades earlier with weighty pros and cons argued by musicologists and performers. Schoenberg, together with Berg and Webern, "solved" the tyranny of harmonic tonality by creating a new musical grammar. This kicked off a half century of controversy—largely among musicians, since the public never abandoned their love of the tonal harmony of their beloved great masters. The battle was joined. Composer/conductor Leonard Bernstein, who wavered at first, soon countered the dark foreboding that all the great music had already been written by composing some very popular works.

This seeking after change for change's sake, concentrating on formulas and other inorganic compositional techniques, did much to confound music's potential in the twentieth century and lose the good graces of the audience. As Prokofiev noted, "There are still so many beautiful things to be said in C major."

More recently, the recording industry cries foul and predicts the end of music if legislation is not adopted to curb the free downloading from the Internet. Downloading music from the Internet is quick, easy, and free; teenagers and others thus flock to it. Of course, the recording industry's worries will soon be over if the happy predictions by some right-wing Christian purveyors of the great apocalypse have their way.

Endism, Theology, and Science

The gloomy epidemic of endism had its emergence long before the twentieth century. If you contemplate an end, then you must logically have a beginning. Thus does theology enter the debate. In the sixteenth century, a former monk, Giordano Bruno, popularized Copernicus's theories of an infinite universe, challenging the idea of creation. He argued that the universe is unlimited in space and time, without beginning or end. Though he considered himself a loyal Catholic, he was immediately arrested. The Pope's theologian saw an effort to subvert the Church, contradicting not only the Church's teaching but all sources of authority. It also shattered the Catholic visions of hell and heaven. Bruno never recanted and in 1600 was burned at the stake. Kepler's revolutionary observations and espousal of the Copernican worldview brought a warning from Cardinal Bellarmine, who said that anything aside from the Bible's description of the sun moving, rising, and setting is heretical. Galileo was also condemned and, placed in house arrest, was forced to recant his scientific beliefs, but the scientific revolution of the sixteenth and seventeenth centuries was unstoppable. The 1642 English social and scientific revolutions put England in the forefront of scientific research, leading to the industrial revolution a century later and eventually to the information revolution, relativity, the computer age, the quantum revolution, and the detritus from all of these "revolutions."

In his last major speech, leading twentieth-century cosmologist Allan Sandage effectively closed the book on endism: "The stars are impossibly old but they are not eternal. They are impossibly far away but not infinitely far away. Stars are born and die. Galaxies are born and die. Atoms are born and die. Particles are born and die. The forces of nature, and perhaps even the dimensions, are born and die." One might well add that the laws of physics and musical forms are born and die. And governmental systems are born and die!

72

Doomsday: The End
of The End

In less than ten years, evangelical minister Timothy LaHaye sold more than 50 million copies of his nine books (in addition to the children's books) in his Left Behind series, based on a strictly literal reading of biblical references to the Second Coming. He received an advance of $45 million from Bantam Dell publishers for his next four novels. A major Hollywood manager, Michael Ovitz of Artists Management Group, signed him up. Warner Books, Bertelsmann, and other publishers rushed in to sign up other evangelical writers.

Apocalyptic LaHaye foresees 144,000 Jews converting to Christianity, thus making possible the Second Coming of Christ, whose thousand-year rule ends in a great battle against Muslims, Jews, and "godless" Christians. Christian music is the fastest growing section of the music business, and Christian films, videos, radio, and events have mass-marketed sites. Rather than condemning popular culture, as was their custom in the past, they have rapidly adopted the popular siren songs, from rock and rap music to tales of romance. Their overflowing cash coffers buy television and radio stations that broadcast twenty-four-hour sophisticated, well-produced programs.

In the mid-seventies, a film about a group of teenagers facing the "Rapture" had a theme song, "I Wish We'd All Been Ready," that became popular in Christian concerts for a decade. In the late nineties, the Internet had sites like *raptureready.com* and *prophecynewswatch.com.*

In seeming contradiction to its beliefs, the Christian right has tightly embraced Israel and the Jews. The founding of Israel is considered the "supersign" that the Second Coming is approaching. Right-wingers Jerry Falwell, Tom DeLay, Ralph Reed, and Oklahoma Republican senator Jim Inhofe are deeply connected to Israeli leaders Ariel Sharon, Benjamin Netanyahu, and the Likud Party. Vigorously pro-Israel organizations such as the American Israel Public Affairs Committee give important support and legitimacy to the Christian right. Some Israeli and American Jews (including this writer/musician) are extremely uneasy about this unlikely embrace with the Christian fundamentalists, given their history of anti-Semitism. While Israel and, to a lesser degree, Jews in general are praised, there is an unmistakable undercurrent of anti-Semitism. And while the political right courts the American Jewish community for their financial and electoral support, some Jews fear that the political right, including a significant minority holding high office, would be lenient to if not supportive of anti-Semitic movements should circumstances favor that stance. Indeed, anti-Semitic incidents, fed by two-faced messages targeting secular Jews, are escalating in many countries.

Pat Buchanan's bestseller, *The Death of the West: How Dying Populations and Immigrant Invasions Imperil Our Country and Civilization,* is pernicious in other ways. Despite his meager showing (0.43 percent of the national vote), he managed to sway the 2000 presidential election by thousands of votes mistakenly cast for him in Florida's heavily Democratic districts. Buchanan's cry of alarm targeted immigrants who "threaten to deconstruct the nation we grew up in and convert America into a conglomeration of peoples with almost nothing in common, not history, heroes, language, culture, faith." I guess that includes my parents and the parents or grandparents of most Americans. America is a land of transplants, and the only persons that have a first claim to the land are indigenous Indians. It is the truth in the lie that seduces; Buchanan further confuses by capturing an element of truth. I refer not to his clever use of the immigration issue, clever because that issue is exploited by rabble-rousing ultraconservatives such as Jean-Marie Le Pen in France and his counterparts in other countries, but because he points up real problems such as the competition for dwindling jobs. He sets one group against another, dividing those who should logically be on the same side.

The truth is that no one group, be it racial, religious, or economic, can be safe from global warming, a great tidal wave, or a nuclear holocaust. These will affect top government officials and corporate CEOs just as they do the lowliest Mexican immigrant for whom the corporate elite has such contempt. No hilltop or cave, no matter how high or how deep, will save the rich or the poor, musicians or scientists.

Endism for Real

The stark symptoms of real endism have already begun to invade the body earth, gaining momentum, yet moving slowly enough so that we have gradually, step-by-step, become accustomed to accepting inconvenience after inconvenience. We boil water and drink bottled water (those who can afford it), peel apples and close our windows to avoid smog. We lose hundreds of species daily. Our fish, beef, and who knows what other foods are toxic.

The deterioration has become serious—poisoning of the earth's water and air, depletion and chemical degradation of the earth's soil, and increasing incidence of breast and other cancers. Autism has increased tenfold in ten years. American mother's milk is ten times as toxic as in European countries. A handpicked group of eleven leading atmospheric American scientists, in a study requested by the Bush White House, con-

cluded that global warming has put us well on the way to a major catastrophe. A vast majority of scientists throughout the world agree. Six gla-
ciers have already melted, and the melting of the polar ice cap could
cause the oceans to rise by up to 160 feet, wiping out the state of Florida
and placing Manhattan under 30 feet of ocean water!

The White House rejected its own study, as they did the Kyoto Pro-
tocol, negotiated by 160 nations (including the U.S.!). The world is being
strong-armed into adapting to all kinds of bio and political engineering,
taking one step at a time into the hell of a toxic planet. These steps become commonplace with little room for deliberation or debate. It all becomes part of the blackening landscape. Recourse to the courts or to the
corporate-controlled media has become less and less possible.

On the flip side, there is a growing backlash. Scientists, musicians, and
a number of people from all walks of life, some for the first time, are taking a leaf from Einstein's example and are joining with political activists.

73

Music and Physics
to the Rescue?

*"To what purpose should I trouble myself in
searching out the secrets of the stars, having
death or slavery continually before my eyes?"*
—ANAXIMENES, sixth century B.C.E.

By now, you the reader know that this book finds
music and physics to be integral to if not representative of the rest of the world, for science and music are
the common language. Like science and music, food,
children, dance, and play are shared by all cultures. Today's wars should have taught us one lesson above all
others: If you disrespect a people's culture, they feel that
you disrespect them. The secret for gaining friendly
communication with an individual or a people is tied up
with one word: respect.

My Arab "enemies," at war with Israel, obviously felt
that I respected their music and culture. Except for individual efforts and people-to-people dialogue, other paths
have hit a dead end. Our foreign policies and long-term
strategies simply have not worked. One tragic lesson after another demonstrated the ineffectuality of the most

powerful military machine in human history. Yet our leaders are insisting on using that same massive power, throwing billions at yet more powerful killing methods. Education, child care, social security, health care, and many other former "priorities" and promises are thrown to the winds. We brag about our success in assassinating a "terrorist" leader or two and ignore the "collateral" killing of civilians (which not infrequently results in the killing of our own soldiers in the process). All this does is invite retaliation and more bloodshed for us as well as the "them" of the moment.

The one way to put a stop to this bloody cycle is to get inside the heads of the people worldwide who do not like us, and find out why. The asinine assumption of some columnists and pundits is that "they" are jealous of our freedoms. No consideration is given to the causes: occupying their lands, killing their people, destroying their families, homes, and infrastructure. By throwing threats, money, or missiles at countries and leaders, we temporarily win "allies," but these "friends" sooner or later prove to be unreliable. Meanwhile, we only sow resentment and lose our own liberties. Cockamamie arguments about why people dislike us convince few except those who wish to delude themselves or are deluded by the media deluge of lies and obfuscation. At this point in history, while we still enjoy overwhelming power, is the precise time when we could use it beneficially. *Honest, truthful* (emphasis needed these days) dialogue with "enemies" (Palestinians, Iraqis, even Muslim leaders) would be a good start.

But above all, we have to address the heart of the matter—the grinding poverty of the majority of the world's people. We must recognize the despair of poor and humiliated people that drives them to commit horrific acts. This will take time, patience, and expenditures of money to alleviate root causes. It will be worth it.

Last Act

We are all cast in this last act of the drama being played out by the human species, a culmination of the hundreds of thousands of years of our human comedy/tragedy. There may never be another chance, so we'd better ponder how we are going to write and direct this act. The denouement—the very fate of our species—is in our hands. We are all the writers, directors, and actors.

As such, we have to learn to listen in depth to the needs and complaints of all peoples, just as a conductor must listen, not only to the first-chair musicians but also to the last stand of the second violins and every other player in the orchestra. We must listen to those voices that are unheard or distorted by simplistic sound bites, media pundits, and politicians. We must even listen to those who have done us harm, the better to counter their lies and peccant plans.

Like John Lennon's song "Imagine," let's imagine what could be if we were to employ our incredibly rich talent in all disciplines to stop the global carnage and turn the world around, exploring better ways of organizing society to give voice to the needs and concerns of all the peoples on the planet and to rid ourselves of obsolescent thinking and functioning. Our long-term plans must focus on finding effective ways of "making music together" with all the peoples of the world, in one great orchestra, performing together as in one vast global concert hall.

A nice thought, but right now the orchestra is in shambles. The peoples of the world must act to defend themselves against those who are dragging us into mass misery and cataclysmic destruction. Is cataclysmic too strong a word? Just ask any group of scientists. A large majority fears that continued corporate malfeasance and man-made environmental damage are approaching the point of no return. Fortunately, people are beginning to realize what their leaders are doing to their families, their world, and the future of their children. The deadly, ultimate race is being run between the forces of destruction and a beautiful new age. Which shall we choose?

Index

About the Author

Music Director of the Symphony for United Nations (SUN) and Principal Guest Conductor of the Central Philharmonic in Beijing, **Joseph Eger** has conducted, guest conducted, or performed with some of the major orchestras of the world in London, Moscow, New York, Pittsburgh, Chicago, Dallas, Boston, Athens, Haifa, Bucharest, China, Greece, Canada, Vienna, Los Angeles, Washington, D.C., and elsewhere. He was Associate Conductor in the U.S. and Europe to Leopold Stokowski. Eger toured worldwide as solo concert artist on the French horn and was labeled by *The New York Times* one of the "greatest French horn players alive." He has produced and conducted numerous innovative events: twelve historic concerts in Carnegie Hall combining, for the first time, symphony and rock music; the first multimedia show; a Lincoln Center concert with John Lennon and Yoko Ono; a unique symphony/rock/jazz concert at the Apollo Theater in Harlem; and other similar combinations. He initiated citywide festivals in New York, Pittsburgh, and elsewhere, winning five mayoral awards. He was one of two men in the world (with Robert Muller, Assistant Secretary General of the United Nations, ret.) who received the Eleanor Roosevelt Man of Vision Award from 500 Women of Vision in the U.S. Capitol. Maestro Eger has taught and given master classes at the Aspen Institute, Peabody Conservatory, New School for Social Research, the Universities of Pennsylvania, and North Carolina, Florida Atlantic University, Nova Southeastern University, and abroad.